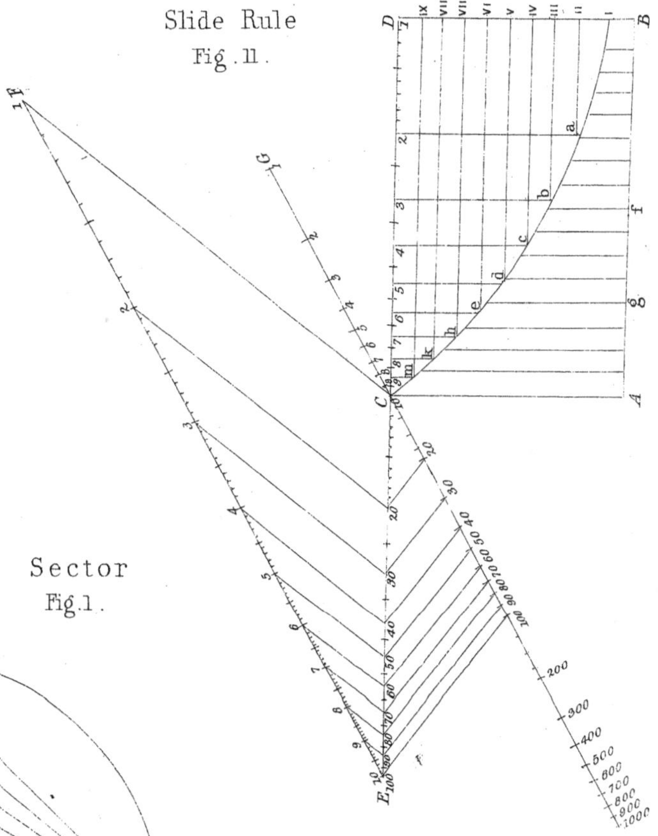

Slide Rule
Fig. II.

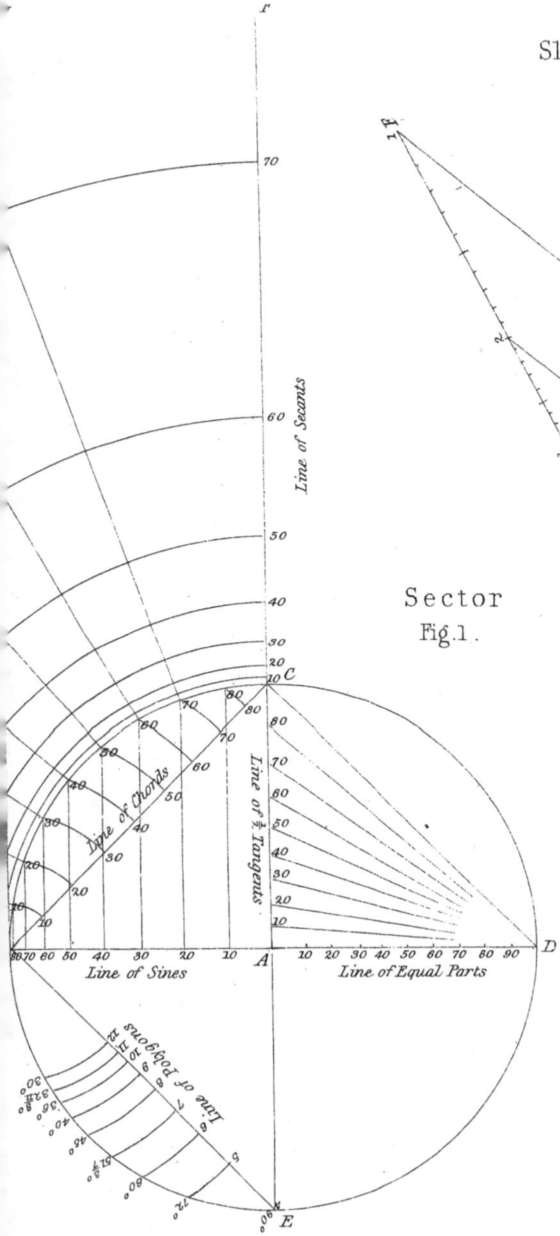

Sector
Fig. 1.

Line of Secants

Line of Chords

Line of ½ Tangents

Line of Sines

Line of Equal Parts

Line of Polygons

A TREATISE

ON A

BOX OF INSTRUMENTS

AND

THE SLIDE-RULE.

FOR THE USE OF GAUGERS, ENGINEERS, SEAMEN, AND STUDENTS.

BY THOMAS KENTISH.

PHILADELPHIA:
HENRY CAREY BAIRD,
SUCCESSOR TO E. L. CAREY.
1852.

Printed by T. K. & P. G Collins.

PREFACE.

In the first edition of this Treatise, its utility, in the absence of other works upon the subject, was assigned as an apology for its publication. The instruments, whose uses it explains, are often so little understood, that scarcely half of them are of any service to their possessor. The Sector, in particular, the most important in the box, is generally regarded as unintelligible. The Slide-rule is briefly noticed in some of the treatises on Mensuration; but, as the pupil is presented merely with a few formal precepts *how* to use it, without knowing *why*, he never understands its nature, never understands the method of determining the real value of any result, and, accordingly, soon lays it by with dissatisfaction, and banishes it from his memory.

The steady sale which the first edition has met with has convinced the Author that his labours were not in vain, and that he has extended among many thousands a knowledge of intrinsic value to all employed in the delineation of mathematical figures.

No attempts, however, are perfect in the beginning; and much was wanted in the first edition to render the work complete. This additional information has been supplied. Several problems have been prefixed, requiring only the compasses and ruler, which, together with those that follow, embrace all that are truly useful, and preclude the necessity of referring to other works on Practical Geometry.

In books upon this subject, it is not usual to annex reasons for any of the operations, but it has been thought advisable to do so, in a few instances, with the more difficult problems; with the rest it is not attempted, because, to have entered fully upon the subject, would have been to transcribe the whole of the Elements of Euclid, a work which is within the reach of every one, and which every one *must* study, who desires thoroughly to understand Geometry.

The part relating to Trigonometry, though concise, will be found to comprehend every thing necessary to enable the student to obtain a clear conception of the subject, and when carried out in connection with the portion devoted to Navigation, will render its acquirement alike easy, pleasing, and useful.

The chapter on Logarithms is written simply to show the mode of adapting them to instrumental computation; a purpose to which every part of the work is, as a matter of course, as much as possible made subservient.

The section relating to the Slide-rule has been entirely re-written; and, in this portion of the work, the Author flatters himself there will be found much that is perfectly new, and many remarks calculated to awaken and stimulate the youthful mind to think for itself; a habit of the utmost value in mathematical science, which, being based on truth, courts investigation, and requires that we shall never assent till we can comprehend. In this part, the formulæ for surfaces and solids have been so modified as to embrace almost every species of mensuration under the simplest form; questions for practice are interspersed throughout, that the student may test his proficiency, and acquire facility in the use of the rule; and tables are inserted at every step, for the purposes of computation; a practice in all cases advisable, as the instrumental operation and numerical calculation necessarily check and illustrate each other.

The reciprocals of divisors, employed as factors, are convenient in practice; but it was deemed advisable, upon the whole, to omit them, as the formulæ for numerical computation would have then been different from those suited to the Slide-rule, which would have tended to perplex the mind of the learner; whereas, by retaining the same form for both operations, it is obvious that to understand one is to understand the other; and the student, instead of coming to regard the instrumental mode of solution as something entirely distinct from the numerical, and looking upon the agreement of the two rather as a coincidence than a consequence, as is too often the case, will see that, in fact, they are identical, and cannot fail, in a short time, of having the very clearest conception of the whole of the subjects treated of.

It is somewhat surprising, that, after the lapse of two hundred years, so excellent an instrument as the Slide-rule should be so little known and appreciated by mathematical students in general. To the engineer and the excise officer it is perfectly familiar, and of daily utility; but, from its having been almost exclusively confined to them, there is an idea prevalent among gentlemen engaged in education, that it can neither be understood by their pupils nor be of any utility to them. A more erroneous conception, on both accounts, cannot be formed; for a knowledge of the instrument is acquired with little or no effort, and it may be truly stated, that it is the most valuable adjunct to mathematical study that can possibly be desired. Nothing imprints a fact so firmly on the mind as repeated exercise. As Demosthenes, when asked the three principal requisites in oratory, summed them up in the word action; so may we say of learning, that the three great essentials to its success are contained in the word repetition. Dexterity in every art, and skill in every science, must be acquired by this means, and by this alone. But, in the solution of questions that are necessarily laborious, every one feels a great disinclination to work through *many* examples, much less to *repeat* them; the consequence is, so little impression is made on the memory, that the knowledge is, in many instances, forgotten as soon as acquired. Now, by the Slide-rule, the most tedious calculations are effected nearly as easily as the most simple. The student can, therefore, after accompanying them the first time with the numerical solution, go over the operations again and again with the rule, with the greatest ease and rapidity, deepening the impression each succeeding time, and rendering the knowledge obtained distinct and permanent.

In the truth of this, the Author is not only borne out by his own experience, but he can refer, with pleasure, to schools in which they have been adopted, and in which they have proved of the greatest assistance; and no one, really fond of knowledge, who may give them a trial, will regret the little extra trouble they may cause, but will rejoice in having found so excellent an aid to study. Mathematical science is of such extensive utility that it ought to be universally understood; and it is impossible to go five or six times through the present work, which, after the first, may be done in a very few days, without being as familiar with the Surfaces and Solids, and with Trigonometry and Navigation, as with the multiplication table; and this is the great object to be attained. To be barely acquainted with them is not sufficient; knowledge, to be useful, must be at the moment accessible, so that we may be enabled to proceed without error or hesitation; and that the most intimate familiarity with the above-mentioned studies will be obtained by the method here pointed out, has been again and again tried, and with the happiest results.

The small section allotted to Land Surveying does not properly come within the design of the work, but it was thought it might prove useful, and has accordingly been inserted. The measuring of a field, which is all that can at all be consistently aimed at here, is so very simple, that one example was deemed sufficient as a guide; but, in teaching the subject, more is necessary; and a very efficient method is to draw on a piece of paper a sketch of a field, which, with the help of a feather-edged plotting scale, or a diagonal scale and a pair of compasses, the pupil should measure, and enter his notes in a field-book, or slate, ruled for the purpose. The sketch should now be handed to the tutor. The learner, then, from his notes, is to construct another, upon paper, from the same scale. When finished, its correctness can be readily ascertained by laying it upon the original, and holding them up to the light, when, if accurately laid down, the lines will, of course, correspond. This plan has been tried for many years, and found to convey a very good idea to the mind of the learner. A little occasional field-practice, which is indispensably requisite, soon renders the study pleasant, and the progress certain.

The chapter on Cask Gauging will, it is humbly hoped, prove a valuable acquisition to the gauger. The great uncertainty and inconvenience of the four varieties render it extremely desirable to have some general, and, at the same time, easy, and easily remembered rule of approximation; and from the method employed in making casks, it is obvious that the exact agreement of their shape with any definite geometrical solid must be perfectly fortuitous. The four varieties, however, are exemplified, together with the general rule for frustums; which latter, though rather tedious, would soon become familiar if once adopted.

In Navigation, for working a day's reckoning, the rule is peculiarly convenient, and sufficient for all practical purposes; superseding the incessant turning over and transcribing from tables; which, though in themselves they are one of the most splendid inventions of all time, and, in elaborate calculations requiring minute exactness, indispensable, are yet, *in their application*, as perfectly mechanical as the instrumental operation itself; so that no reasonable objection can be urged against the adoption of the Gunter, that does not apply, with equal force, to the use of Logarithms altogether.

For gentlemen, however, who may not desire to use the Slide-Rule, it may be here stated, that the work by no means absolutely requires it; it is equally available as a Treatise on Mensuration, Trigonometry, and Navigation. For the purposes of calculation, it would be found a great convenience to copy out, upon a sheet of Bristol board, the tables at pages 115, 116, 118, 123, 126, 129, 136, 137, 138, 150, 182, and 198, as it would save much needless turning over of the pages; and if each were enclosed in borders, and slightly washed over with different colours, it would make them of easier reference.

In studying Trigonometry, Wallace's Practical

Mathematician's Pocket Guide will be found a convenient set of Logarithmic Tables; their cost is a mere trifle. Barlow's and Galbraith's Tables are extremely useful. The latter contains the secants and cosecants, which, as complemental to the cosines and sines, offer great facilities in calculation.

CONTENTS.

A TREATISE

ON

A BOX OF INSTRUMENTS.

A BOX OF INSTRUMENTS.

THE contents of a case of Mathematical Instruments are, generally, a pair of plain compasses, a pair of bow compasses, a pair of drawing compasses, and a drawing pen; a parallel ruler, a protractor, a plain scale, and a sector. The plain compasses consist of two inflexible rods of brass, revolving upon an axis at the vertex, and furnished with steel points. The bow compasses are a smaller pair, provided with a pen for describing small circles in ink. The sides of the pen are opened or closed with an adjusting screw, that the line may be drawn fine or coarse as required. The drawing compasses are the largest of the three; one of the legs is furnished with a socket for the reception of either of the four following pieces, as occasion may require:—1. A steel point, which, being fixed in the socket, makes the compasses a plain pair, like the other; 2. A port-crayon, for the purpose of carrying a piece of blacklead, or slate-pencil, according as paper or slate is used for drawing upon; 3. A steel pen, like the one attached to the bow compasses, but larger, for the purpose of describing circles of greater diameter; 4. A rowel, or spur wheel, with a brass pen above it, for the reception of ink, which the spur, in its circuit, distributes in dots upon the paper. The pen, the port-crayon, and the dotting wheel, are each furnished with a joint, that, when fixed in the compasses, they may be set perpendicular to the paper. The drawing pen is the same as the steel pen of the compasses, only that it is screwed upon a brass rod, of a convenient length for the hand, and into the rod itself is inserted a fine steel point for pricking.

The parallel ruler consists of two flat pieces of ebony or ivory, connected together by brass bars, having their extremities equidistant, by which contrivance, when the ruler is opened, the sides necessarily move in parallel lines. The protractor is a semicircular piece of brass, divided into 180 degrees, and numbered each way, from end to end. In some boxes this is omitted, and the degrees are transferred to the border of the plain scale. The plain scale is a flat piece of box or ivory, and is so called from containing a number of lines divided into plain or equal parts. A scale of chords, of a fixed radius, is also graduated upon it. The sector is a foot rule, divided into equal portions, movable upon a brass joint, or axis, from the centre of which are drawn various lines through the whole length of the ruler. The legs represent the radii of a circle, and the middle of the joint expresses the centre. The lines upon the sector are of two sorts, single and double: the single lines run along the margin and the edges; the double lines radiate from the centre to the extremities of the legs, and are marked twice upon the same face of the instrument, in order that distances may be taken upon them crosswise, when they are opened to an angular position.

PRACTICAL GEOMETRY.

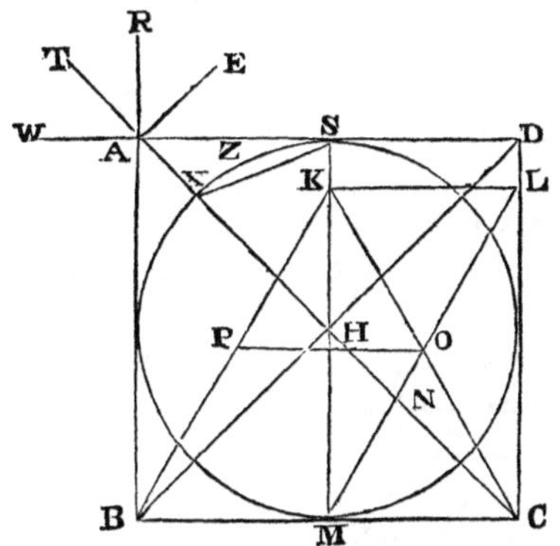

DEFINITIONS.

A POINT is that which has position, without length, breadth, or thickness.

A line is length, without breadth, or thickness.

A superficies is length and breadth, without thickness.

A solid is that which has length, breadth, and thickness.

An angle is the opening of two straight lines meeting in a point, as RAE.

Lines which run side by side, and are always equidistant, are called parallels, as SD, KL.

A line is perpendicular to another when the angles on both sides of it are equal; and each of these angles is called a right angle; thus RA is perpendicular to WS; and the angles RAW, RAS, are right angles.

An acute angle is less than a right angle, as EAS.

An obtuse angle is greater than a right angle, as TAS.

A figure containing three sides is called a triangle.

The boundary of a right-lined figure is termed its perimeter.

A triangular figure containing three equal sides is an equilateral triangle, as KBC.

If two of its sides only are equal, it is an isosceles triangle, as BHC, in which BH = HC.

If the three sides are unequal, it is a scalene triangle, as KBH.

A triangle containing a right angle is called a right-angled triangle, as ABC.

A triangle containing an obtuse angle is an obtuse-angled triangle, as KHB.

An acute-angled triangle contains three acute angles, as NMC.

A figure containing four sides is called a quadrilateral.

A parallelogram is a quadrilateral whose opposite sides are parallel, as SKLD, KPOL.

A rectangle is a parallelogram whose angles are right angles, as KMCL.

A square is a rectangle whose sides are equal, as ABCD.

A rhomboid has its opposite sides equal, but two of its angles are obtuse, and two acute, and these are opposite to each other, and equal, each to each, as KBML.

A rhombus, like a square, has all its sides equal, but two of its angles are obtuse, and two acute, and these are opposite to each other, and equal, each to each, as PBMO.

When a quadrilateral has only one pair of its sides parallel, it is called a trapezoid, as PBCO.

When none of its sides are parallel it is a trapezium, as KHNL.

A line crossing a quadrilateral from opposite angles, is termed a diagonal; thus AC, BD, are the diagonals of the square ABCD.

Figures of more sides than four are called polygons.

If all the sides and angles are equal, it is a regular polygon; if unequal, an irregular polygon.

A polygon of five sides is termed a pentagon; of six, a hexagon; of seven, a heptagon; of eight, an octagon; of nine, a nonagon; of ten, a decagon; of eleven, an undecagon; and of twelve, a dodecagon.

A triangle is sometimes called a trigon; and a quadrilateral, a tetragon.

A circle is a plane figure bounded by a curved line called the circumference, which is everywhere equidistant from the centre.

A right line passing through the centre, and meeting the circumference at each extremity, is called the diameter, as SM.

A right line reaching from the centre to the circumference is termed the radius, as HM.

An arc of a circle is any part of the circumference, as the curve from S to X.

A chord is a right line joining the extremities of an arc, as the straight line from S to X.

A segment is a space included between an arc and its chord, as SZXS.

A sector is a part of a circle contained by two radii and the arc between them, as SHXZ.

Hence a sector is made up of a triangle and a segment.

A semicircle is half a circle; a quadrant, the fourth part; a sextant, the sixth part; and an octant, the eighth part.

A rectilineal solid whose ends are equal, similar, and parallel, and whose sides are parallelograms, is called a prism.

If the ends also of the prism are parallelograms, it is a parallelopiped: if all the sides are square, it is a cube: if the ends are unequal and dissimilar, it is a prismoid.

A cylinder is a round solid, of uniform thickness, having circular ends.

A pyramid is a solid which has a rectilineal base, and triangular sides meeting in a point called the vertex.

A cone is a round solid tapering uniformly to a point.

A sphere is a solid every way round.

A segment of a solid is the part cut off the top by a plane parallel to its base.

A frustum is the part left at the bottom, after the segment has been cut off.

Prisms, cylinders, pyramids, and cones are said to be right or oblique according as the base is cut perpendicularly or obliquely to the axis.

Plain figures formed by the cutting of a cone are called conic sections. A cone may be cut five ways. If the cutting plane passes through the vertex of a right cone and any part of the base, the section is an isosceles triangle; if through the sides, parallel to the base, a circle; if obliquely through the sides, an ellipse; if through one side and parallel to the other, a parabola; if in any other way, the cutting plane will run beyond into a similar cone inverted over the other, and then the section is termed an hyperbola. (*N. B. An ellipse may be considered as an elongated circle, and may be described by driving in two pins as centres, passing over them a string with a loop at each end, and working a pencil round within the string, keeping it stretched to its limits.*) The centres are usually called foci: the diameter passing through them is termed the transverse; the short one at right angles to it, the conjugate diameter.

A line perpendicular to either of the diameters is called an ordinate; and the sections of the diameter met by it are termed abscissas.

The vertex of a conic section is the point where the cutting plane meets the opposite sides of the cone. The axis of a parabola or hyperbola is a right line drawn from the vertex to the middle of the base.

All round solids may be conceived to be described by the rotation of planes on their sides, or diameters, as axes.

A right-angled triangle rotating on its perpendicular, forms a cone; a parallelogram, revolving on its side, produces a cylinder; if a circle turn upon its diameter, it shapes out a sphere; and the revolution of an ellipse generates a spheroid. If the ellipse revolves on the transverse diameter, the spheroid is called prolate; if on the conjugate, an oblate spheroid. The figure formed by the revolution of a parabola about its axis is termed a paraboloid, or parabolic conoid; the solid formed in the same way by an hyperbola, an hyperboloid, or hyperbolic concoid.

If a section of a curve revolve on a double ordinate as axis, it will generate a spindle; and this will be circular, elliptic, parabolic, or hyperbolic, according to the curve from which the section is taken.

A regular body is a solid contained under a certain

number of similar and equal plane figures. There are but five kinds, which are, the tetraedron, having four equal triangular faces; the hexaedron, or cube, which has six equal square faces; the octaedron, which has eight equal triangular faces; the dodecaedron, which has twelve equal pentagonal faces; and the icosaedron, which has twenty equal triangular faces.

THE COMPASSES.

1. To bisect a given line AB.

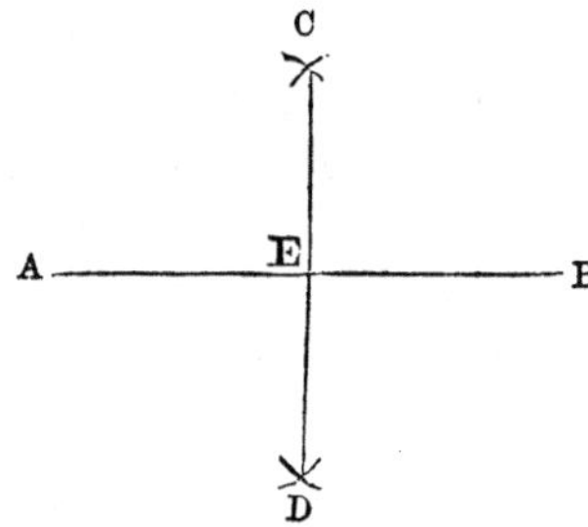

From A and B as centres, with any radius greater than half AB, describe arcs cutting each other in C and D. Join the points C and D, by drawing the straight line CD; this will be perpendicular to AB, which it will bisect in the point E.

2. To bisect a given angle ABC. (See p. 21.)

From B as a centre, with any radius, describe the arc DE. From D and E, with the same, or any other radius, draw arcs cutting each other in F. Join BF, and it will bisect the angle as required.

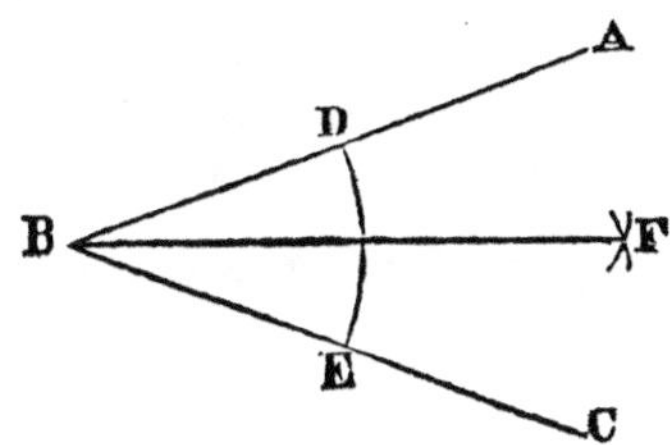

3. To erect a perpendicular to a straight line AB, from a given point C within it.

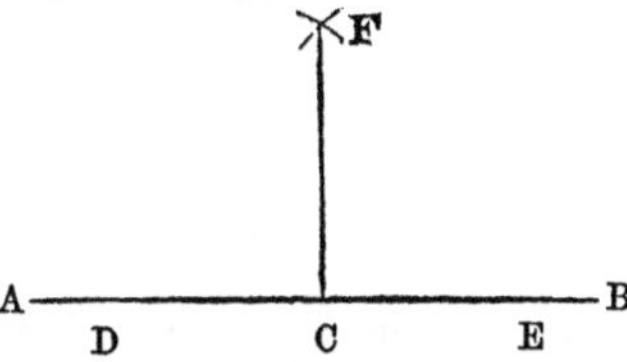

When the point C is near the middle of the line, on each side of it set off any two equal distances CD, CE. From D and E as centres, with any radius greater than CE or CD, describe arcs cutting each other in F. Join FC, and it will be perpendicular to AB.

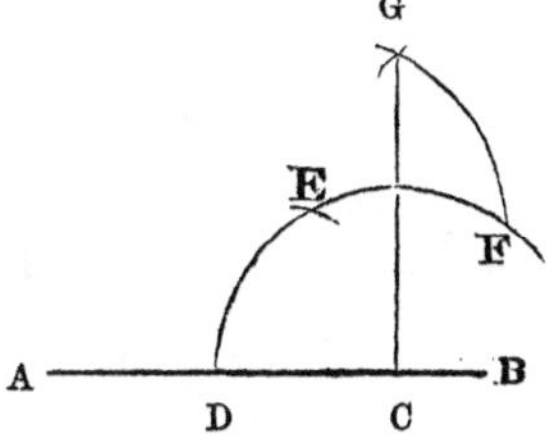

When the point C is at or near the end of the line, from C, with any radius, describe the arc DEF. From D, with the same radius, cross it in E. From E, with the same radius, describe the arc GF; and from F, with the same radius, cross the last arc at G. Lraw GC, and it will be perpendicular to AB.

4. To draw a perpendicular to a line AB, from a point C without it.

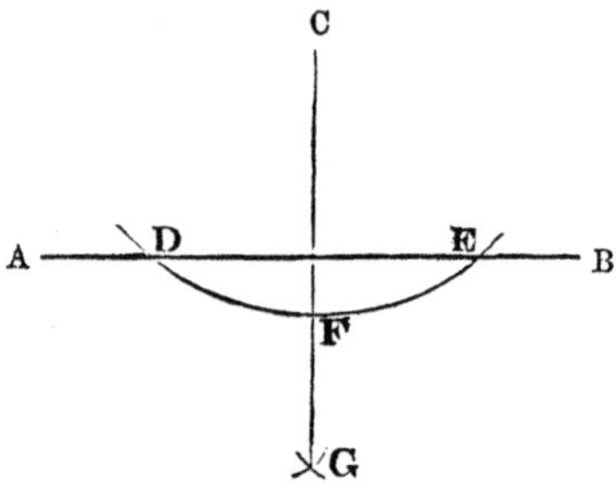

When the point C is nearly opposite the middle of the line, from C, with any convenient radius, describe the arc DFE crossing AB in D and E. From D and E, with the same or any other radius, describe arcs cutting each other in G. Draw CG, and it will be perpendicular to AB.

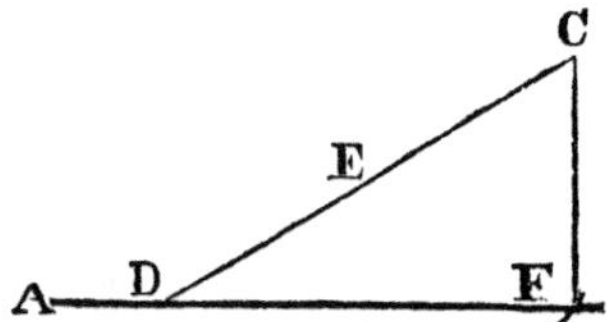

When the point C is nearly opposite the end of the line, from C draw any line CD. Bisect CD in E, and from E, with the radius EC, cross AB in F. Draw CF, and it will be perpendicular to AB.

5. To make an angle equal to a given angle ABC.

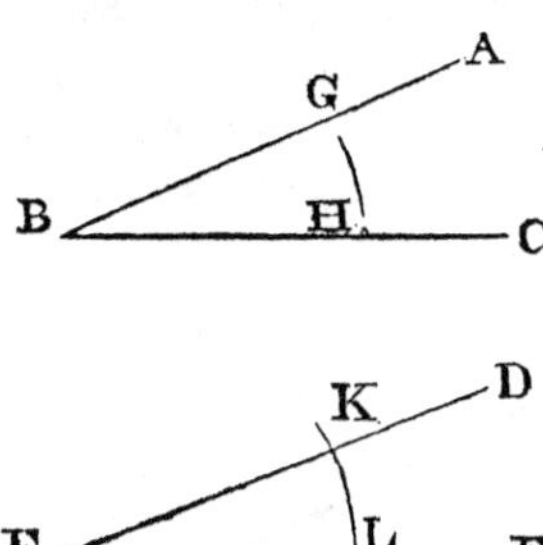

From B, with any radius, draw the arc GH; and from E, with the same radius, describe the arc KL. Make KL equal to GH, and through K draw the straight line ED. The ∠ DEF = ∠ ABC.

6. To describe a circle through three given points A, B, and C.

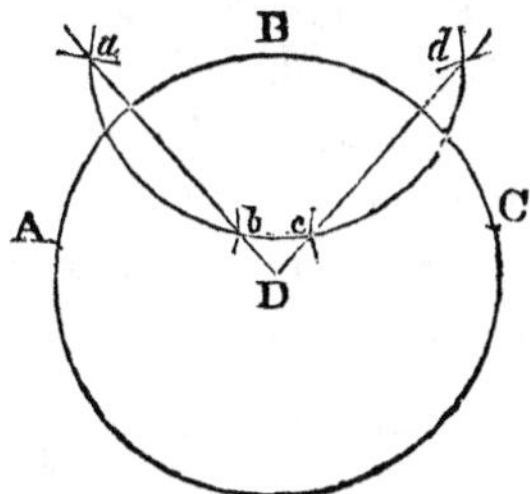

From B, with any radius less than BA, describe the arc *abcd;* and from A and C, with the same radius, cross it in *a* and *b*, *c* and *d*. Draw straight lines through the points of intersection to meet in D, which will be the centre of the circle required.

To find the centre of a given circle, take any three points in the circumference, and proceed in like manner.

To describe a circle about a triangle, select the three angular points, and proceed in like manner.

7. To construct a triangle of which the three sides A, B, and C are given.

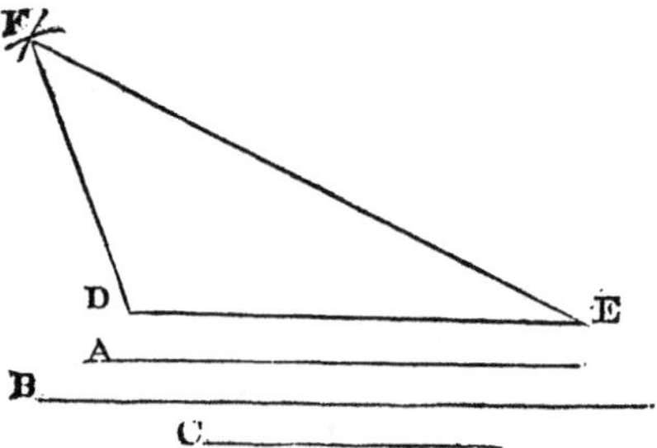

Draw the line DE equal to A. From D, with C for a radius, describe an arc at F; and from E, with B for a radius, cross it at F. Draw the lines DF, EF, and DEF will be the triangle required.

To construct an equilateral triangle proceed in the same manner, taking the base each time as radius.

8. To construct a rectangle, whose length A B and breadth C are given.

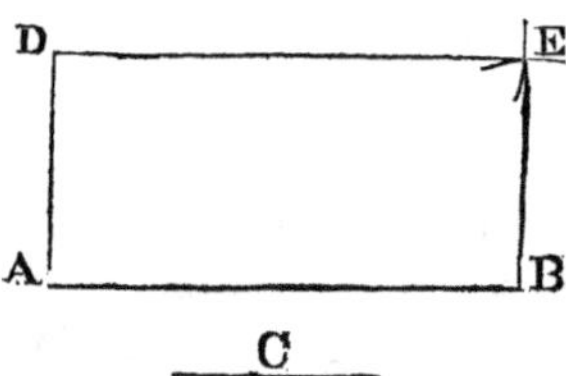

At A erect a perpendicular AD, equal to C. From D,

with the distance AB, describe an arc at E; and from B, with the radius C, cross it at E. Draw DE, EB. ADEB is the parallelogram required.

To construct a square, make the perpendicular equal to the base, and proceed in like manner.

To construct a rhombus and rhomboid, determine the angle by problem 5, and then proceed in like manner.

9. To draw a line parallel to a given line AB, at a given distance.

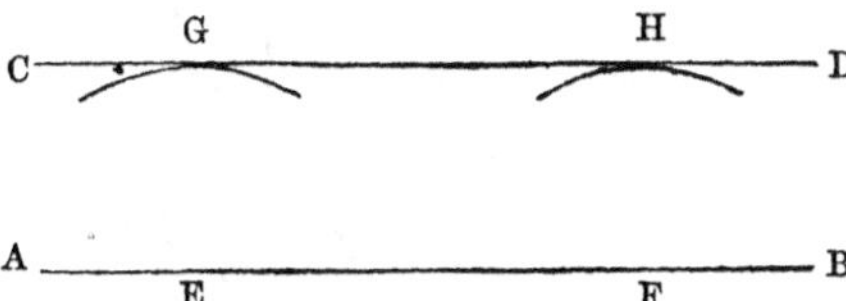

Take any two points E and F in the line AB, and, with the given distance, describe the arcs G and H. Draw the line CD touching the arcs, and it will be parallel to AB.

When the line is to pass through a given point C.

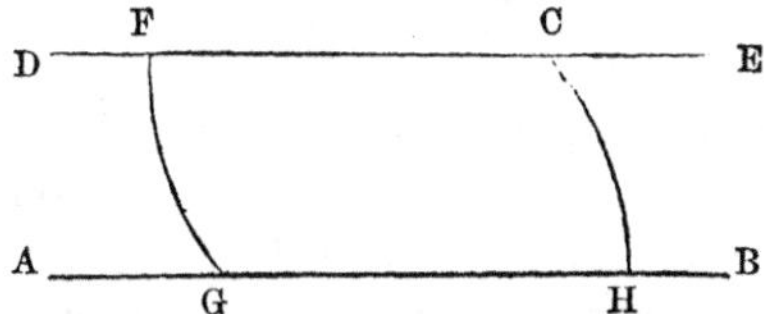

In AB take any point G, and with the distance GC describe the arc CH; from C, with the same radius, describe the arc GF, and make FG equal to CH. Through F and C draw the straight line DE, and it will be parallel to AB, as required.

10. To project an ellipse, or oval, the length A B and breadth A C, being given.

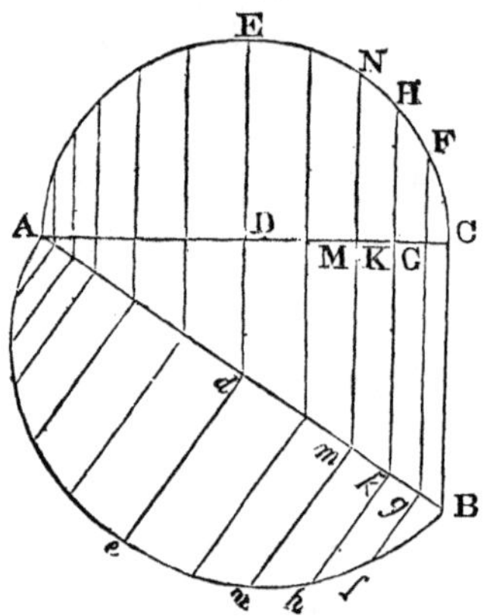

Bisect AC in D, and upon it describe the semicircle AEC. At C draw a straight line CB, perpendicular to AC; and from the point A, with the given length as a radius, cross CB in B. From the semicircle AEC and parallel to CB, draw any number of straight lines FG*g*, HK*k*, NM*m*, ED*d*, &c. On the line AB, at the points of intersection, *m*, *k*, *g*, &c. erect perpendiculars, and make *gf* equal to GF, *kh* equal to KH, *mn* equal to MN, &c. Lastly, trace a curve line from B through the points *f*, *h*, *n*, *e*, &c., and it will give half of the ellipse, from which the other half may readily be constructed. This method is of great utility in describing elliptical arches, staircases, &c., and for any purpose in which circular figures, or figures of any shape, require to be elongated without altering the breadth, as in cutting gores for globes, &c.,

in which case AB will be equal to half the circumference, the breadth DE being regulated by the number of gores employed. In instances like these, the length AB will be very great compared with the breadth DE, and AB may then be divided in the same proportion as AC, by other means; for example, if the length is to be 12 times the breadth, then each of the distances B*g*, *gk*, *km*, will be 12 times the corresponding distances CG, GK, KM. This method of projecting ellipses is derived from conceiving a right cylinder to be cut by two planes, one parallel, and the other oblique to the base.

THE PARALLEL RULER.

1. Through a given point A, to draw a line parallel to a given line DE.

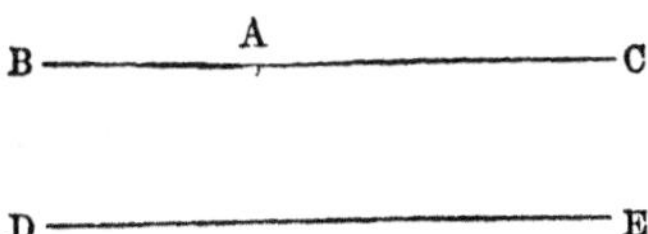

Lay the edge of the ruler upon DE, and move it upwards till it reaches the point A, through which draw BC. BC is parallel to DE.

2. To make an angle equal to a given angle ABC.

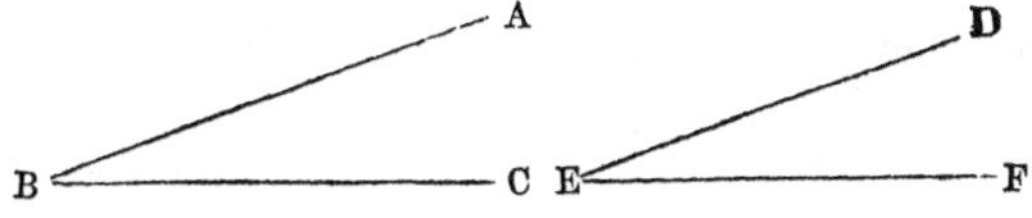

Lay the base EF, in a line with BC, and draw ED, parallel to BA.

The $\angle$ DEF $= \angle$ ABC.

3. To find a third proportional to two given straight lines, AB, AC.

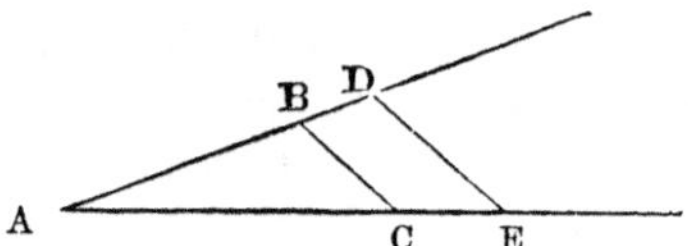

Place them together so as to form any angle DAE. Take AD = AC. Draw BC, and DE, parallel to it.

AB : AC : : AC : AE.

4. To find a fourth proportional to three given straight lines, AB, CD, EF.

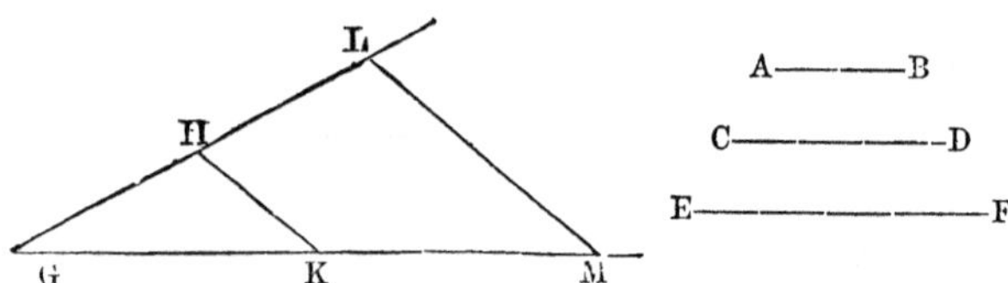

Make any angle, LGM. Take GH = AB, GK = CD, and GL = EF. Join HK, and draw LM parallel to it.

AB : CD : : EF : GM.

For GH : GK : : GL : GM.

5. Another method. Given AB : CD : : EF : ?

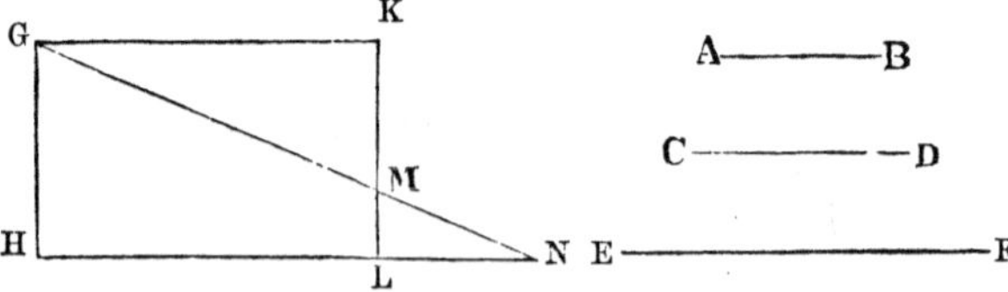

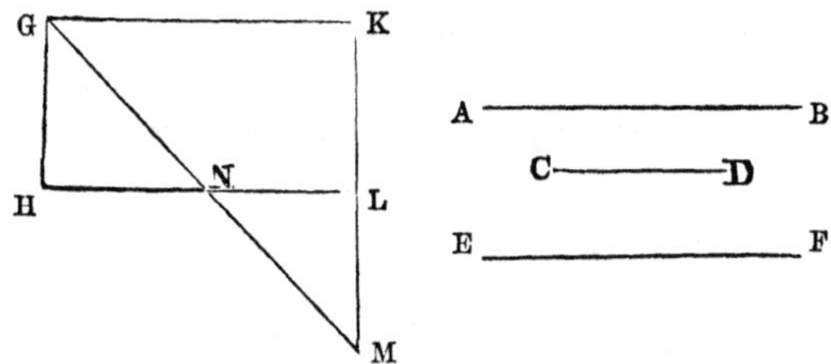

Make a rectangle GHLK, of the second and third, CD, EF; and in one of the sides, produced if necessary, take KM, equal to the first, AB. Draw GM, to meet HL, produced if necessary, in N.

AB : CD : : EF : HN.

For KM : GH : : KG : HN.

And the third problem may be performed in a similar manner, by making a square of the second term.

6. Hence to inscribe a square in a given triangle ABC.

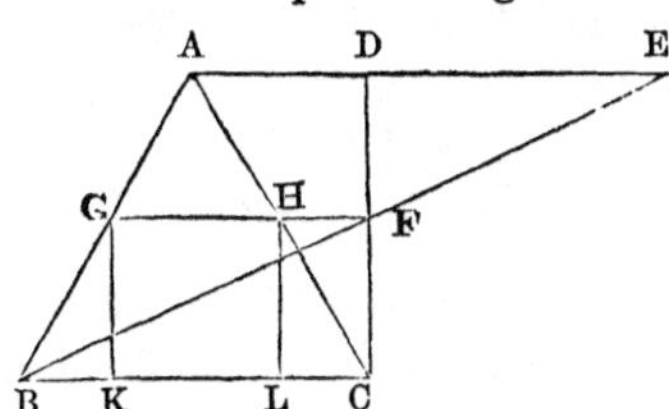

Through the vertex A, parallel to BC, draw the straight line AE, and from C raise a perpendicular to meet it in D. Draw DE, equal to DC. Join EB, cutting DC in F. Through F draw FG, parallel to BC, and through H and G draw HL, GK, parallel to DC. GL will be the square required.

For FC : CB : : CD : BC + DE

That is, GH : CB : : CD : BC + CD $\therefore$ GH $= \dfrac{CB \,.\, CD}{BC + CD}$.

That is, the side of the inscribed square is equal to the product of the base and altitude divided by their sum: or the sum is a fourth proportional to the side, base, and altitude.

7. To divide a given line AB, similarly to another CD.

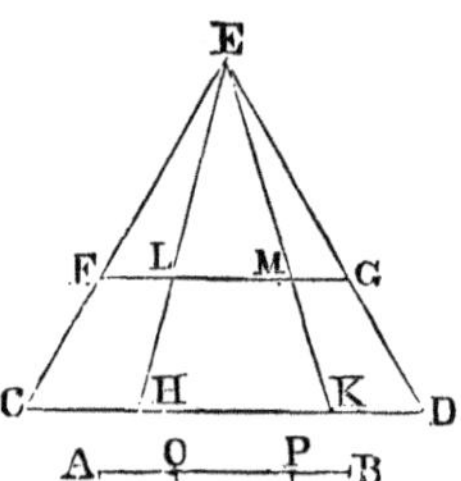

On CD, construct the equilateral triangle CDE, and from the vertex downwards cut off EF = AB. Draw FG, parallel to CD, and join EH, EK. Transfer the divisions L and M, to O and P. AB is divided, in the points OP, similarly to CD, in H and K, that is,

CD : AB : : CK : AP : : CH : AO.

8. Another method. Let AB be the divided line, and AC the line required to be similarly divided.

Lay them together, making any angle CAB. Join the extremities CB, and draw GE, FD, parallel to CB. AC is divided similarly to AB; that is,

AB : AC : : AE : AG : : AD : AF.

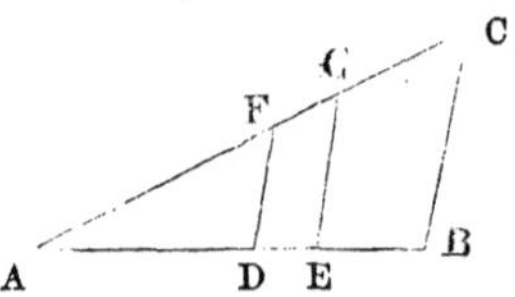

9. Hence to divide a line AB, into any number of equal parts, as four.

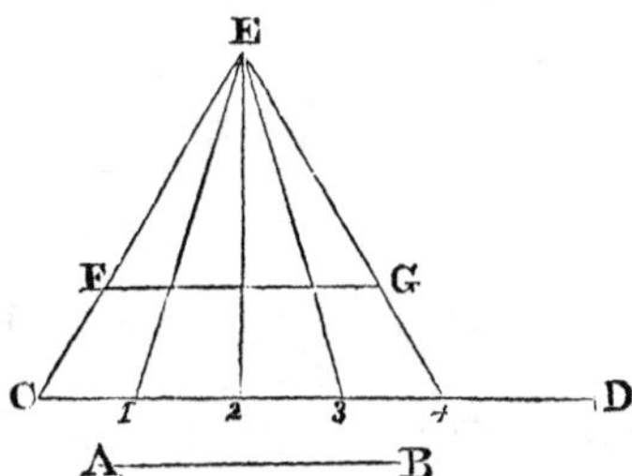

Draw an indefinite line CD, and from C, set off any distance the intended number of times, in the points 1, 2, 3, 4. On C 4, construct the equilateral triangle EC 4. Make EF = AB, and draw FG, parallel to CD. FG is equal to AB. Join E 1, E 2, E 3; and FG, that is AB, will be divided into four equal parts.

10. Or by the other method.

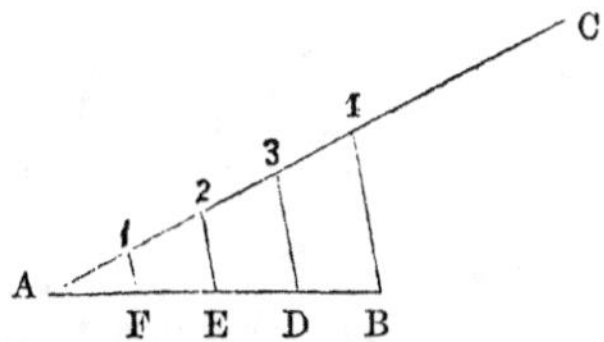

From A, draw the straight line AC, making any angle CAB. From A, set off any distance the intended number of times toward C, in the points 1, 2, 3, 4. Draw the line 4 B, and parallel to it 3 D, 2 E, 1 F. AB will be divided equally into the required number of parts.

11. To reduce a trapezium ABCD, to a triangle.

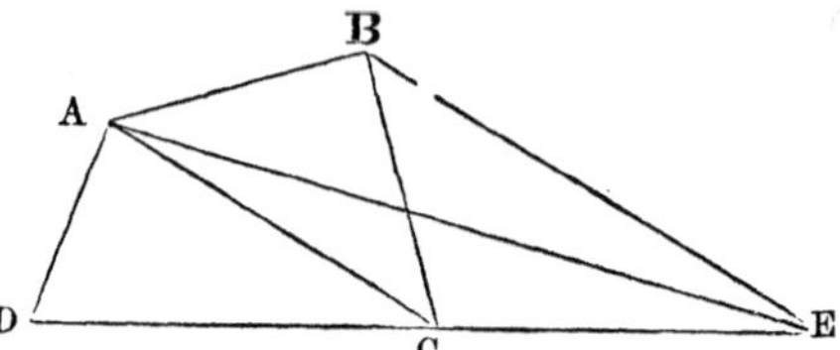

Draw the diagonal AC, and through B draw BE, parallel to it, meeting DC, produced to E. Join AE. The triangle ADE, is equal to the trapezium ABC.

12. To reduce any polygon ABCDE, to a triangle.

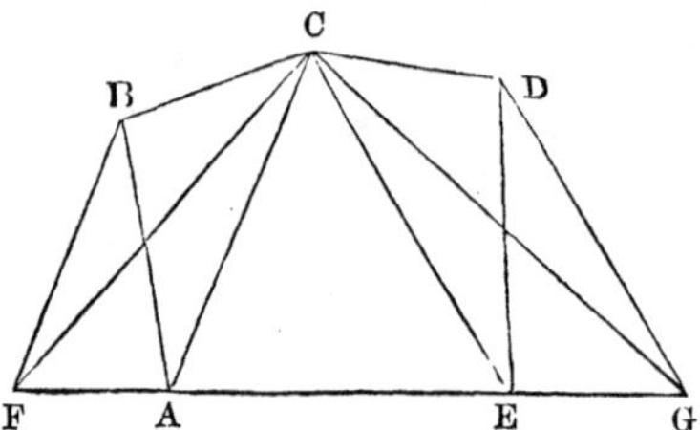

Draw the diagonals CE, CA, and produce AE, both ways, to F and G. Draw DG, parallel to CE, and BF, to CA. Join CF, CG. The triangle CFG, is equal to the polygon ABCDE.

13. To reduce a triangle ABC, to a parallelogram.

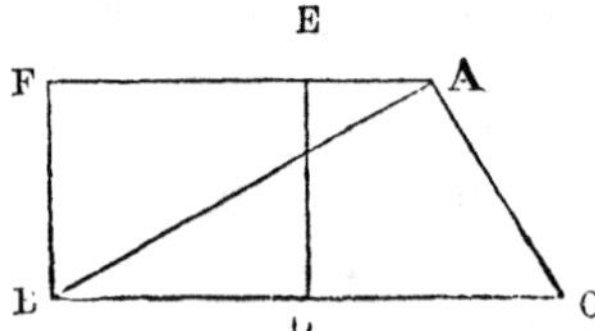

Through the vertex A draw FA, parallel to BC. Bisect BC in D, and raise DE, perpendicular to BC, meeting FA in E. Draw FB parallel to ED. The rectangle FD, is equal to the triangle ABC, for the content of a triangle is equal to the product of half the base by the altitude.

14. To reduce a parallelogram ABCD, to a square.

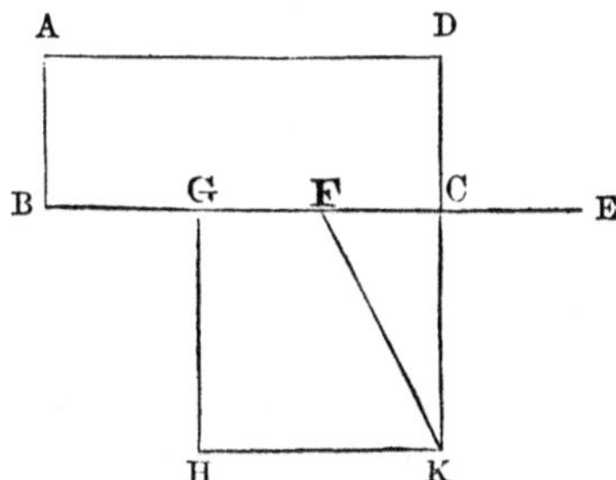

Produce BC to E, making CE = DC. Bisect BE in F. Produce DC to K, making FK = FE, and CG = CK. Through K draw KH, parallel to BE, and through G draw GH, parallel to DK. The square GK, is equal to the rectangle AC, and CG is a mean proportional between DC and CB; that is,

$$DC : CG :: CG : CB.$$

15. Hence to make a square equal to any given polygon ABDC.

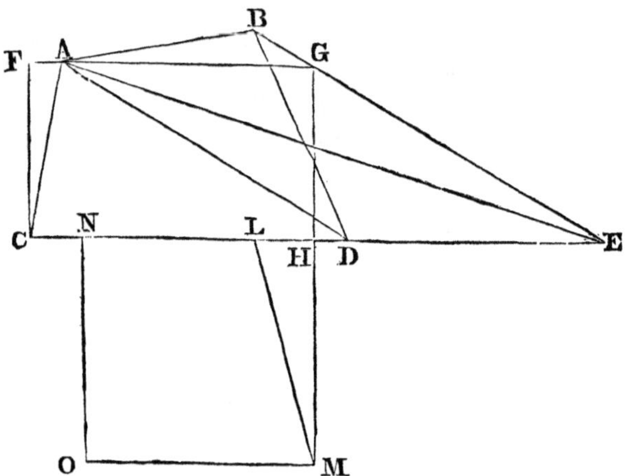

Reduce it to the triangle ACE, and this to the parallelogram CHGF, and this to the square NM.

16. To reduce a triangle ABC, to another that shall be of a given altitude AD.

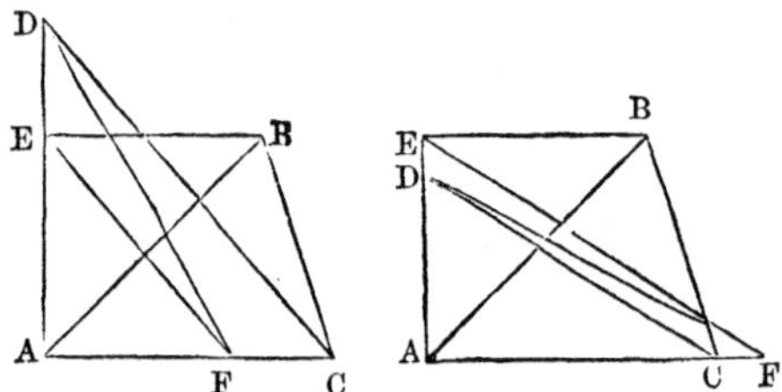

Join DC, and through the vertex B draw BE, parallel to the base AC, meeting AD, produced if necessary, in the point E. Through E draw EF, parallel to DC. Join DF. The triangle DAF = ABC.

THE PROTRACTOR.

It is unnecessary to describe the construction of this.—It is simply a semicircle, divided into 180 equal parts, termed degrees. As mentioned in the introduction, these degrees are, in some boxes, transferred to the border of the plain scale, which is used precisely as the protractor: it is, however, far from being so convenient as the semicircle. Some protractors are complete circles, and contain, of course 360 degrees.

USES.

1. To find the number of degrees contained in any given angle BAC.

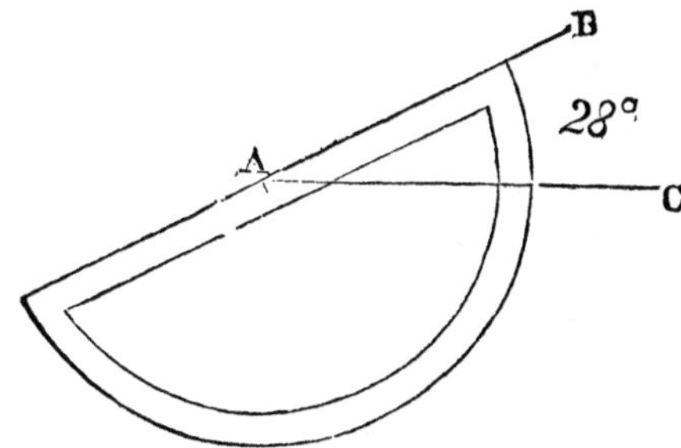

Lay the central notch of the instrument upon A, and the edge along AB, as in the diagram; and observe the number cut by the other line AC.

2. To lay down an angle ABC, which shall contain a given number of degrees.

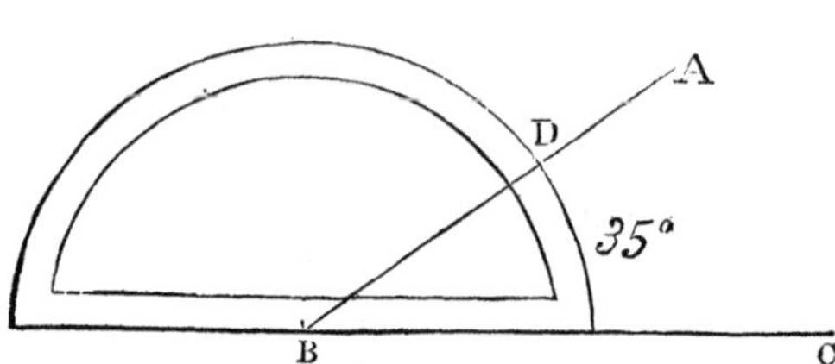

Draw a line BC, of any length. Place the notch against B, and the edge along BC. Prick a point D against the number required, and through it draw the line AB.

3. Through a given point C, to draw a line paralell to a given line AB.

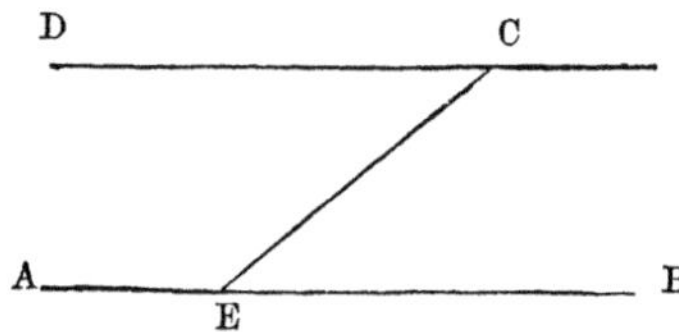

In AB take any point E, and join CE; and make the angle DCE, equal to the angle CEB, by the line DC. DC is parallel to AB.

4. To divide a given line AB, into any number of equal parts.

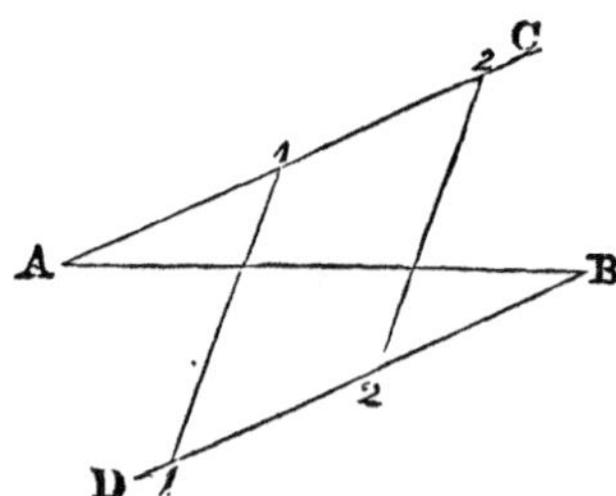

From the extremities of the line AB, draw the lines AC, BD, making equal angles. From A towards C, and from B towards D, set off any distance once less than the intended number of parts. Number one from the extremity A, and the other towards the other extremity B, and join the like numbers. AB will be divided as required.

5. To erect a perpendicular to a given line AB, from a point C, within it.

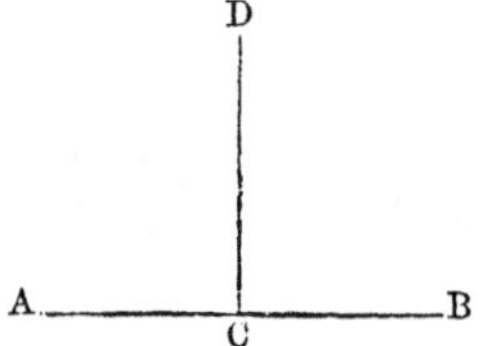

Place the edge along A B, and the notch at C. Then against 90 prick the point D, and draw DC. DC will be the perpendicular required.

6. To let fall a perpendicular upon a straight line AB from a point C.

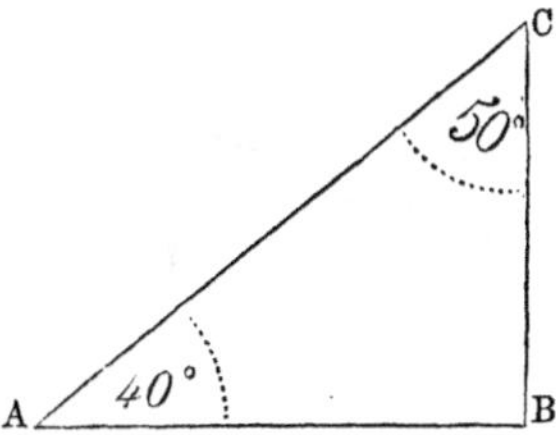

Draw any line CA. Observe the number of degrees contained in the angle CAB. Subtract it from 90, and make the angle ACB, equal to the remainder, by the line CB. CB will be the perpendicular required.

7. To divide a given angle, ABC, into any number of equal parts.

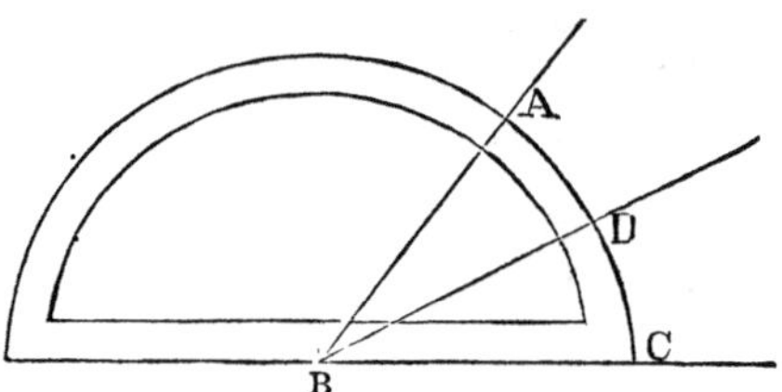

Find the number of degrees; divide it by the required number of parts, and prick off the quotient along the rim, as in D. Join BD.

8. To inscribe a circle in a given triangle, ABC.

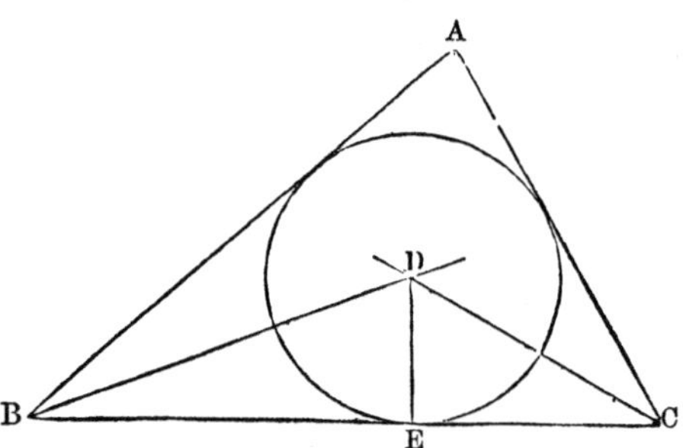

Bisect any two of the angles ABC, ACB, by the straight lines BD, CD, crossing each other in D, from which let fall the perpendicular DE. DE is the radius of the required circle.

9. In a circle to inscribe any regular polygon.

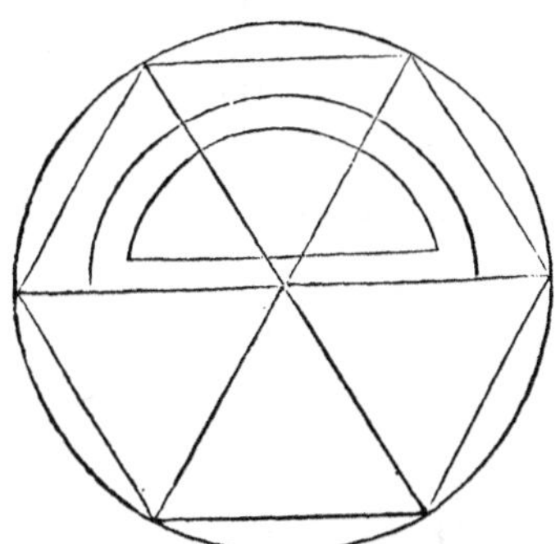

Divide 360 by the intended number of sides. Place the instrument with the notch against the centre, and prick off the quotient round the rim. If the number of sides be odd, the instrument will require to be turned round; if even, half may be pricked off, and lines drawn through the centre, the extremities of which meeting the circle, will give the points required. Connect the points, and the polygon will be completed.

10. To construct a regular polygon on a given line, AB.

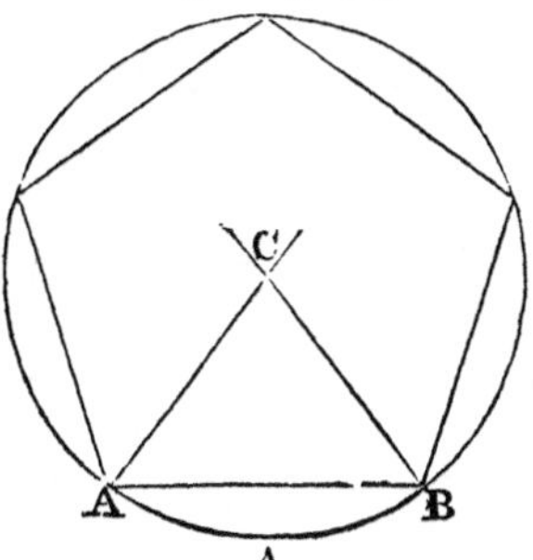

Divide 360 by the intended number of sides, subtract the quotient from 180, and halve the remainder. Make the angles CAB, CBA, each equal to this, by lines intersecting in C. CA is the radius of a circle, round which the line AB may be carried the required number of times.

11. To describe a circle within or without a regular polygon.

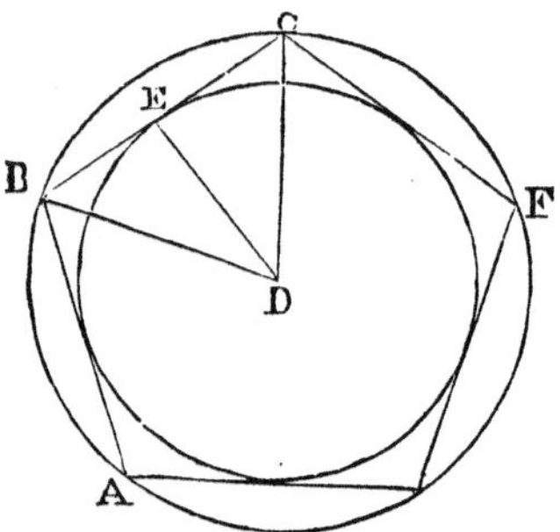

Bisect any two angles, ABC, BCF, by the lines BD, CD, and from D let fall DE, perpendicular to the side BC. BD is the radius of the outer, and ED of the inner circle, as required.*

THE PLAIN SCALE.

THE method of constructing the plain scale is obvious. A number of horizontal lines on one side having been drawn through the whole length of the rule, and a vertical column on the left for the reception of numbers, a distance of two inches is laid down, and divided, upon the top line

* Polygons are more expeditiously constructed by means of the Sector, of which hereafter.

into 9 equal parts, upon the next into 8, and so on, down to 3. These are repeated along the rule as often as its length will allow; the first portion of each is subdivided into tenths and twelfths; and numbers are placed in the column, showing how many of the tenths are contained in an inch. A scale of chords, marked C, is graduated along the top; and at 60 a small *r* is placed, indicating that distance from the beginning to be the radius of the circle from which they are taken. They are merely the degrees of a quadrant or quarter of a circle, laid down in a straight line; thus,—

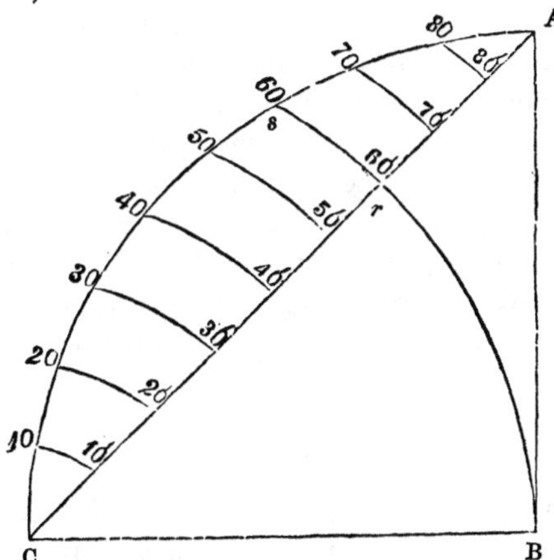

Draw the lines BA, BC, at right angles to each other; and with any convenient distance, BA, describe the arc A*s*C; divide it into 90 equal parts, and join AC. From C as a centre, with the distances C 10, C 20, &c., describe the arcs 10, 10′; 20, 20′, &c. meeting the line AC. Fill up the separate degrees, which are not marked in the diagram to prevent confusion, and the scale is complete.

It is evident, by inspection, that the chord of 60 is equal to the radius, as shown by the letter *r* upon the rule; which distance is therefore always to be taken in laying down angles, as will be described when we come to speak of its uses.

The other face of the rule is divided along the top into inches, and these into tenths. The next line is divided into 50 equal parts. Under these is what is called the diagonal scale. It consists of 11 equidistant parallel lines, crossed by vertical ones a quarter of an inch apart. By taking these alternately, another scale is obtained, of twice the size of the former. The first of each of these is divided into ten equal parts, above and below; and oblique lines are drawn from the first perpendicular below to the first division above, and continued parallel, by which contrivance the first space is divided into 100 equal parts: consequently, if the line contain ten of the large divisions, each of these small spaces is the thousandth part of such line. If, therefore, the large divisions denote hundreds, the first subdivisions will be tens, and the second, units.

USES.

The plain scale is simply for laying off distances. The manner of using the first side is evident. If the number 47 be required, place one foot of the compasses upon 4, and extend the other to the 7th subdivision of the tenths. If 3 feet 5 inches be required, place one foot on 3, and stretch the other to the 5th division of the twelfths. The following figure is laid down from the scale at the bottom, numbered 15, and may be tested by the pupil.

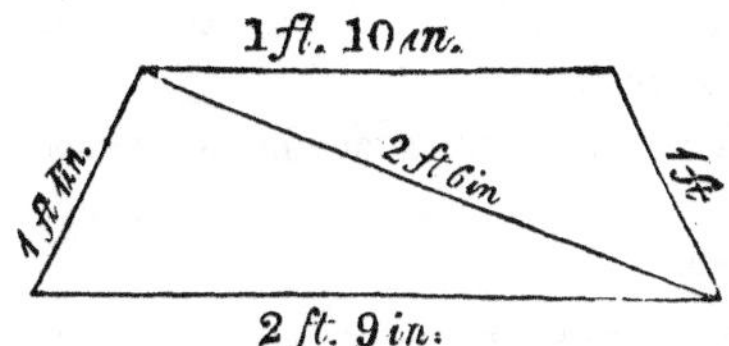

The scale of chords serves the purposes of the protractor, and is used as follows :—

1. To find the number of degrees contained in a given angle, BAC.

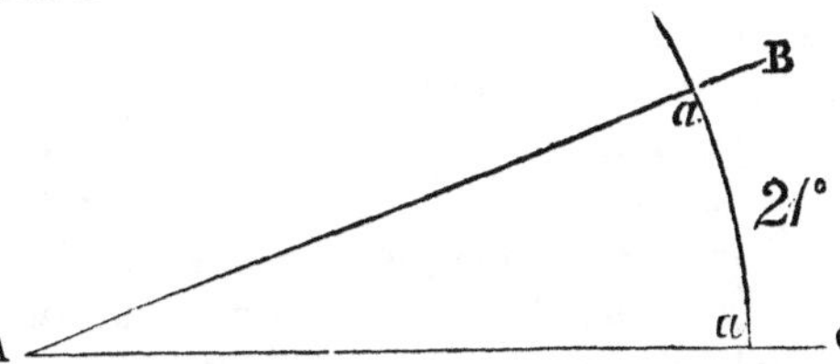

From A with the radius 60 describe the arc *a a*. Take the distance from *a* to *a* in the compasses, and apply it to the beginning of the scale. The number to which it reaches, shows the degrees contained in the angle.

2. To make an angle ABC, which shall contain a given number of degrees, as 26.

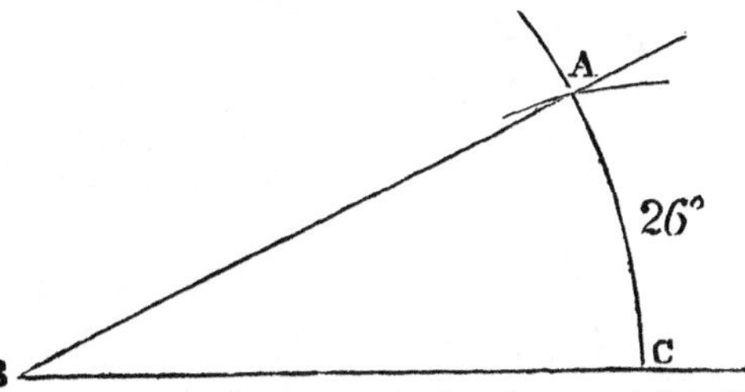

From B with the radius 60 describe the arc AC. Take 26 from the scale; place one foot of the compasses in C, and with the other cross the arc in A. Join AB. ABC is the angle required.

And in the same manner may be performed all the problems described under the protractor, which need not be repeated here.

The diagonal scale is used where more exactness is required, or when a number containing three figures is wanted, as 357, 35.7, 3.57, &c. If the primary divisions denote hundreds, the subdivisions express tens, while the units are counted on the parallels—upwards on the left, the quarter-inch scale; and downwards on the right, the half-inch scale. If the primary divisions (those denoted by the perpendiculars) express tens, the diagonals will be units, and the parallels tenths, and so on; each smaller division being the tenth of the next larger.

In taking off numbers from this scale, proceed in an inverse order to the figures; that is, commence with the units, proceed to the tens, and end with the hundreds; thus, to take off 346 from the larger scale. Place one foot of the compasses upon the sixth parallel where it is crossed by the fourth diagonal, and extend the other to the third perpendicular. To take off 1839 from the smaller scale. On the 9th parallel, where it is crossed by the third diagonal, place one foot of the compasses, and extend the other to the 18th perpendicular.

To raise a perpendicular to a given line, AB.

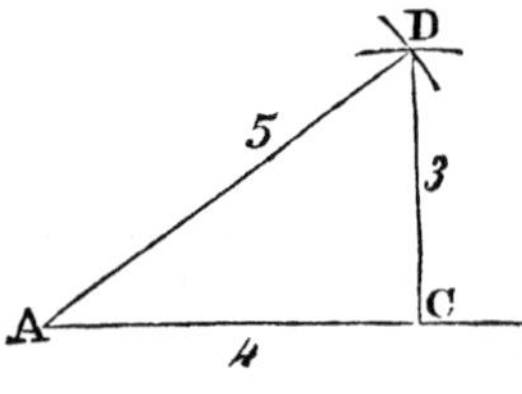

Make AC = 4 equal parts. From C, with a distance of 3 from the same scale, make an arc; and from A with 5 cross it in D. Join DC. DC will be the perpendicular required.

Given two square pieces of deal board; the side of one 17 inches, of the other 29. It is required to ascertain the side of another that shall be equal to both.

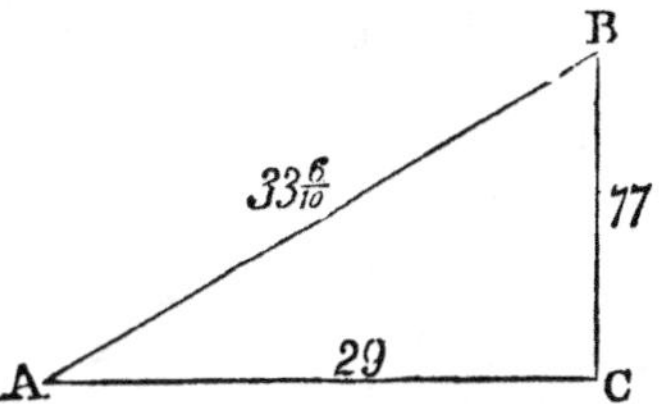

Lay down a base, AC = 29, and raise a perpendicular BC = 17. Join AB; apply it to the scale, and it will be found 33.6. For the square of the hypothenuse is equal to the sum of the squares of the base and perpendicular.

It is required to find the diameter of a copper, that, being of the same depth as another whose width is 13 inches, may contain thrice as much.

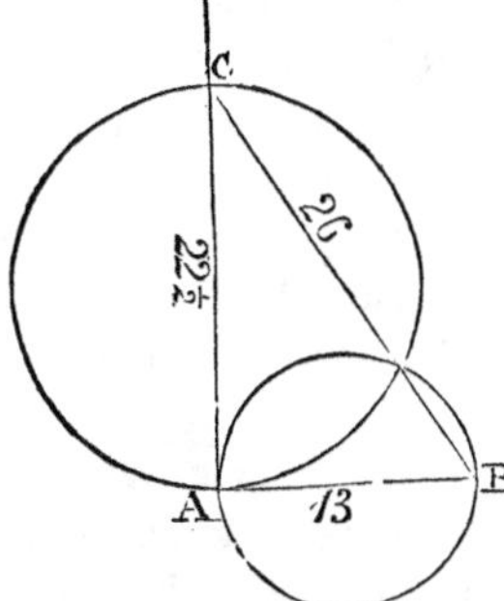

Make AB = 13, and raise a perpendicular AC. From B, with twice the distance, cross it in C. Apply CA to the scale; it will be found to be 22½. For if AB = 1, and BC = 2, then AC = $\sqrt{3}$.

Three men bought a grindstone, 20 inches in diameter; and agreed that the first should use it till he ground down ⅓ of it for his share; the second to do the same; and the third to finish the remainder. Ascertain the diameters of the second and third shares.

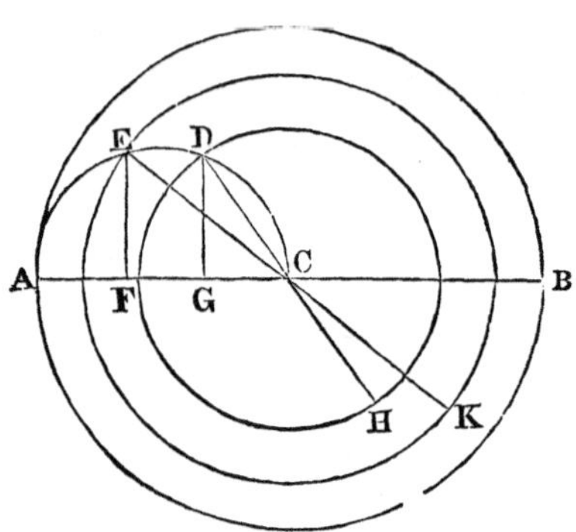

Draw the line AB = 20; bisect it in C, and on AC describe the semicircle AEDC. Divide AC into three equal parts, in the points FG; perpendicular to which draw the lines EF, DG, to meet the semicircle. Join EC, DC, and produce them till CH be equal to CD, and CK to CE. EK is the diameter for the second person, and DH for the third. By applying them to the scale, EK will be found to be about $16\frac{1}{3}$, and DH rather more than $11\frac{1}{2}$. . For $AC.CG = CD^2$ and $AC.CF = CE^2$ $\therefore CG : CF :: CD^2 : CE^2$; and the areas of circles are as the squares of their radii.

Four men bought a grindstone of 30 inches in diameter; and agreed that the first should use it till he ground down 1-4th of it for his share, deducting 6 inches in the middle for waste; and then that the second should use it till he ground down 1-4th part; and so on. What part of the diameter must each grind down?

1-5th of the diameter being waste, 1-25th of the content

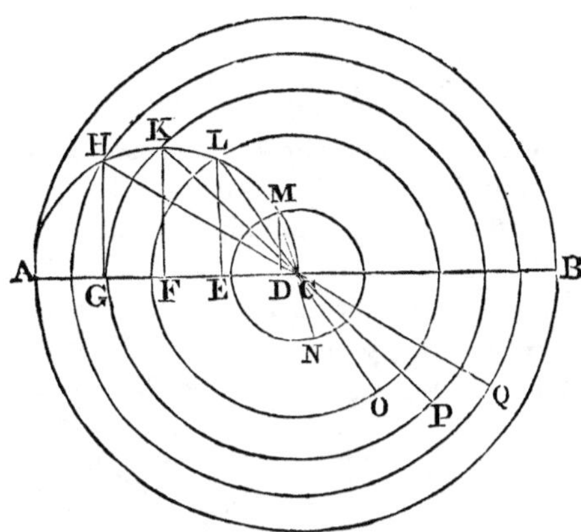

is waste; therefore, conceiving the whole to contain 25 shares, 1 share will be for waste, and each person will have 6 shares.

Hence, draw the line AB = 30, and on AC describe the semicircle AHKLMC. Divide AC into 25 equal parts by problem 10, parallel ruler; and make AG, GF, FE, ED, each equal to 6 of these parts. From the points G, F, E, D, raise perpendiculars to meet the semicircle in H, K, L, M; join HC, KC, LC, and MC; and, having drawn circles from H, K, L, M, with C as a centre, produce them to N, O, P, Q. The diameter AB is 30: HQ will be found, upon applying it to the scale, to measure about 26; KP, nearly 21½; LO, about 16; and MN, 6 for the waste. Subtracting these in succession, we have 4 inches for the first, 4½ for the second, 5½ for the third, and 10 for the last.

The student will find these problems repeated at the end of the Uses of the Slide Rule.

TRIGONOMETRY.

THE circumference of a circle is supposed to be divided into 360 equal parts, called degrees. Each degree is divided into 60 minutes; each minute into 60 seconds; and so on.

Degrees are marked with a small ° at the top of the figure, minutes with ′, seconds with ″, and so on. Thus, 36° 18′ 25″—36 degrees, 18 minutes, 25 seconds.

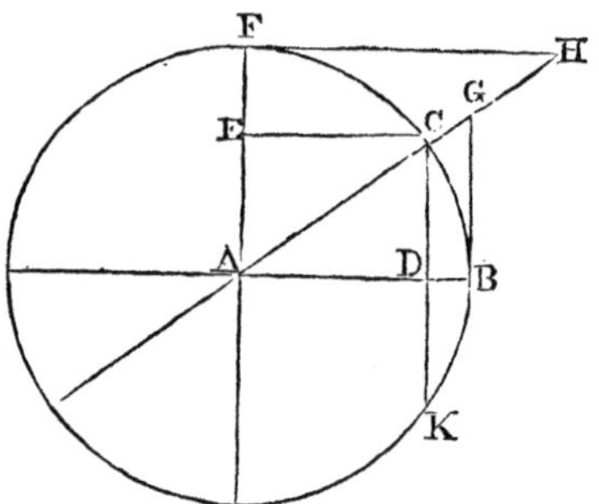

The difference of an arc from 90 degrees, or a quarter, is called its complement; thus FC is the complement of CB. The chord of an arc is a line drawn from one extremity of the arc to the other; thus CK is the chord of the arc CBK.

The sine of an arc is a line drawn from one extremity of the arc perpendicular to the diameter passing through the other extremity; thus CD is the sine of the arc CB, or angle CAB, which it measures; and CE is the sine of the arc CF, or angle CAF, which it measures.

The sine is half the chord of twice the arc, or angle.

The tangent of an arc is a line touching the circle in one extremity of the arc, and meeting a line drawn from the centre through the other extremity; thus GB is the tangent of the arc CB, or angle CAB; and FH is the tangent of the arc CF, or angle CAF.

The secant of an arc is the line meeting the tangent; thus GA is the secant of the arc CB, or angle CAB; and AH is the secant of the arc CF, or angle CAF.

The versed sine of an arc is the part of the diameter intercepted between the arc and its sine: thus DB is the versed sine of CB.

The cosine, cotangent, and cosecant of an arc, are the complement's sine, tangent, and secant: *co* being simply a contraction of *complement.* Thus CE, or AD, is the cosine of CB, being the sine of the complement CF: so FH is the cotangent of the arc CB, being the tangent of the complement FC; and AH is the cosecant of CB, being the secant of the complement CF.

From these definitions it is evident—

1st. That when the arc is 0, the sine and tangent are 0, but the secant is then the radius AB.

2d. When the arc is a quadrant, FB, then the sine is the greatest it can be, being the radius of the circle; and the tangent and secant are infinite.

3d. The versed sine and cosine together make up the radius.

4th. The radius AB, the tangent BG, and secant AG, form a right-angled triangle.

5th. The cosine AD, the sine DC, and radius AC, also form a right-angled triangle.

6th. The radius AF, the cotangent FH, and cosecant AH, also form a right-angled triangle. And since the angle FAH = the angle ACD = the angle AGB; these right-angled triangles are similar to each other. Hence

	AD	:	DC	: :	AB	:	BG
viz.	cosine	:	sine	: :	radius	:	tangent.
	AE	:	EC	: :	AF	:	FH
viz.	sine	:	cosine	: :	radius	:	cotangent.
	AD	:	AC	: :	AB	:	AG.
viz.	cosine	:	radius	: :	radius	:	secant.
	EA	:	AC	: :	FA	:	AH.
viz.	sine	:	radius	: :	radius	:	cosecant.
	GB	:	BA	: :	AF	:	FH.
viz.	tangent	:	radius	: :	radius	:	cotangent.

So the radius is a mean proportional between the cosine and secant, the sine and cosecant, and the tangent and cotangent.

In every case in trigonometry three parts must be given to find the other three; and one of these, at least, must be a side.

The cases in trigonometry are of three varieties:

1st. When a side and its opposite angle are given.

2d. When the two sides and the contained angle are given.

3d. When the three sides are given.

I.

When a side and its opposite angle are two of the given parts; then

Any side : sine of its opp. angle : : any other side : sine of its opp. angle.

And,

Sine of any ∠ : its opp. side : : sine any other ∠ : its opp. side.

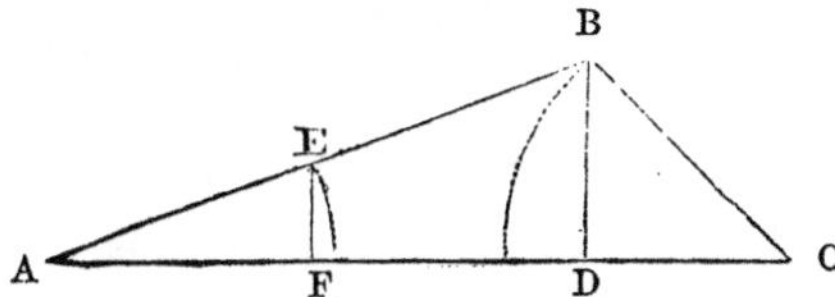

Make AE = BC, and EF, BD, perpendicular to AC.
Then AB : BD : : AE : EF. But AE = BC;
∴ AB : BD : : BC : EF.
But BD is the sine of C, and EF of A;
∴ AB : sin opp. ∠ C : : BC : sin. its opp. ∠ A.

II.

When two sides and their contained angle are given.

Sum of sides : their difference : : tang. ½ sum of opposite angles : tang. of ½ their diff.

Then ½ sum + ½ diff. = greater;
and ½ sum — ½ diff. = less.

Let ABC be the proposed triangle, having the two given sides AC, BC, including the given angle C.

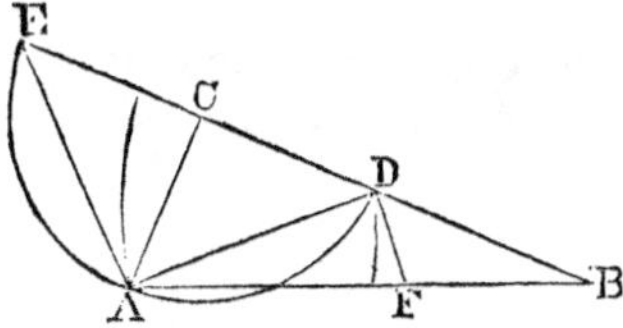

From C, with radius CA, describe a semicircle, meeting

BC produced in D and E. Join AE, AD, and draw DF parallel to AE.

Then BE = BC + CA, the sum of the sides BC, CA: and BD = BC — CA, their difference.

Also, CAB + CBA = CAD + CDA = 2CDA;

$$\therefore \angle \text{ CDA} = \frac{1}{2}\left\{ \text{CAB} + \text{CBA.} \right\}$$

That is, CDA is half the sum of the unknown angles.

$$\text{Again, } \angle \text{ CDA} = \angle \text{ DBA} + \angle \text{ DAB};$$

$$\therefore \angle \text{ DAB} = \angle \text{ CDA} - \angle \text{ DBA}$$

$$= \angle \text{ CAD} - \angle \text{ CBA},$$

Add to each side DAB; then,

$$2 \text{ DAB} = \text{CAD} + \text{DAB} - \text{CBA}$$

$$= \text{CAB} - \text{CBA};$$

$$\therefore \text{DAB} = \frac{1}{2}\left\{ \text{CAB} - \text{CBA.} \right\}$$

That is, DAB is half the difference of the unknown angles.

Now EAD being a semicircle, EA is perpendicular to AD, as is also DF; ∴ AE is the tangent of CDA, and DF the tangent of DAB, to the same radius AD.

But, BE : BD : : AE : DF;

that is, the sum of the sides : difference of the sides : : tangent of ½ sum of opposite angles : tangent of ½ their difference.

Three angles of a triangle being equal to 180°, the sum of the unknown angles is found by taking the given angle from 180°.

III.

When the three sides are given, to find the angles.

Base : sum of other sides : : difference of those sides : difference of segments of base, made by perpendicular falling from the vertex.

Let ABD be the given triangle.

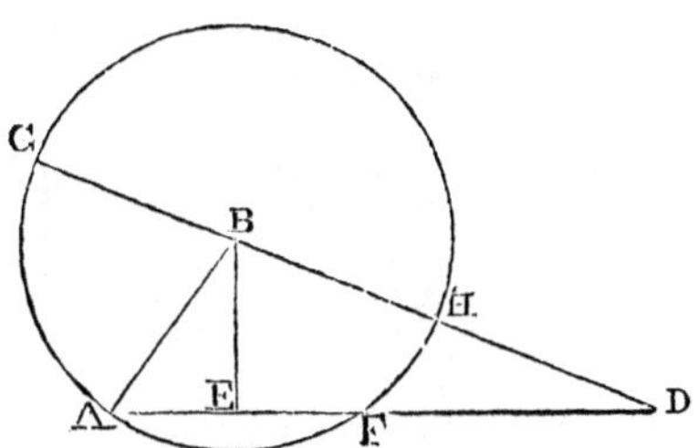

From B with the distance BA describe a circle, and produce DB to G. Then,

DG = DB + BA; and HD = DB — BA.

Also, since AE = EF; ∴ FD = DE — EA.

But AD : DG : : HD : FD, (Euclid 3, 46;) that is, Base : sum of other sides : : diff. of those sides : diff. of segments of base.

Hence, in each of the two right-angled triagles,. there will be known two sides, and the right angle opposite to one of them.

All cases of plane triangles may be solved by these three problems; but for right-angled triangles, the following are more convenient:—

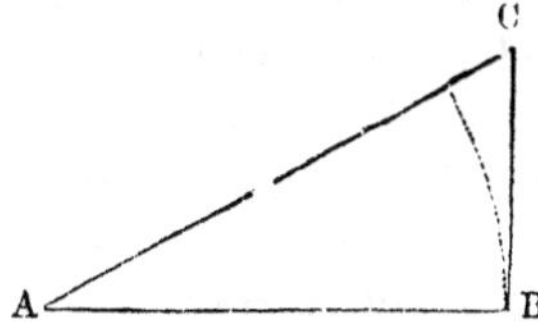

First make the base AB radius; then,

Radius : tang. A : : AB : BC; and
Radius : secant A : : AB : AC.

Next make CB radius.

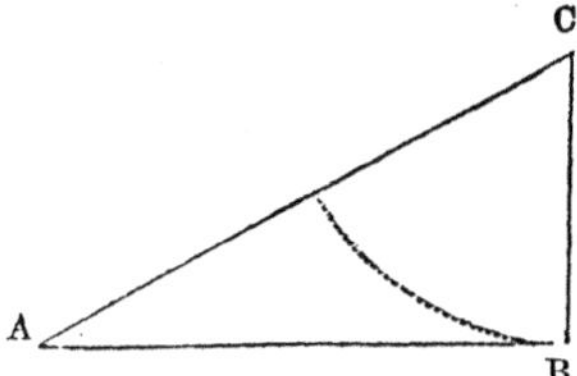

Then Radius : tang. C : : CB : BA; and
Radius : secant C : : CB : CA.

Lastly, make AC radius.

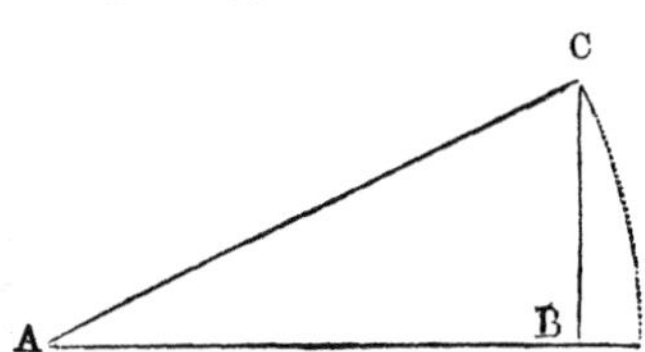

Then, Radius : sine A : : AC : CB; and
Radius : cos. A : : AC : AB.

SECTOR.

THE lines on the sector, as before stated, are of two sorts, single and double. Only the double lines are sectoral; these proceed from the centre, and are—

1st. A scale of equal parts, marked L, and containing 100 divisions.

2d. A scale of chords, marked C, running to 60.

3d. A line of secants, marked S, running to 75.

4th. A line of polygons, marked POL. These are numbered backwards from 4 to 12.

Upon the other face, the sectoral lines are,—

1st. A line of sines, marked S, running to 90. The sines may be easily distinguished from the secants, though marked the same; as the distances of the sines diminish towards the end, while the secants increase.

2d. A line of tangents, marked T, running to 45.

3d. Between the line of tangents and sines there is another line of tangents, beginning at a quarter of the length of the former, to supply their defect, and extending from 45 to 75, marked t or T.

The distance from T to T, from S to S, from C to C, and from L to L, is the same; so that at whatever distance the sector may be opened, the angles formed by those lines will always be equal. The polygons are laid down to a shorter radius, for the sake of including the pentagon and square. The radius, the chord of 60°, the sine of 90°, the tangent of 45°, the secant of 0°, all are equal.

The method of constructing the sectoral lines is exhibited in Fig. 1, fronting the title-page.

From the point A with any convenient radius describe the circle BCDE, and draw the diameters BD, CE, crossing each other at right angles in the centre. Produce C to F, and through B draw BG, a tangent to AB. Join EB, BC, CD.

In the construction of the following Scales, only the primary divisions are drawn, the smaller ones being omitted to prevent confusion:—

Divide AD into ten equal parts, and these again into tenths; so shall AD be a line of equal parts.

Divide the arc BC into 9 equal parts, and these again into tenths, and with the compasses from B, as a centre, transfer the divisions to the straight line BC; so shall BC be a line of chords.

From each of the divisions of the arc BC let fall perpendiculars upon AB, and number them backwards; so shall AB be a line of sines.

From the point A, through the divisions of the arc BC, draw straight lines to meet BG; so shall BG be a line of tangents. And from D to the same divisions of the arc BC, draw straight lines cutting AC; so shall AC be a line of half-tangents.

The distances from the centre A to the divisions on the line of tangents being transferred to AF, AF will be a line of secants.

From B, in the arc BE, cut off the fifth part of the circumference, 72°, and transfer it with the compasses to the straight line BE. Do the same with the sixth, seventh, eighth, &c.; so shall BE be a line of polygons. Transfer these distances to the lines upon the rule, and the sectoral part is complete.*

From the property of similar triangles, AB : BC : : A*b* : *bc*; hence, if AB were divided into ten equal parts, and A*b* contained four of those parts, then BC being divided into ten equal parts, *bc* would contain four of those parts.

And if AB were the sine of 90°, and A*b* the sine of

---

* These lines are best laid down by the help of tables of natural sines, tangents, &c., in the same way as the lines of the slide-rule are laid down by logarithmic numbers, sines, and tangents; of which hereafter.

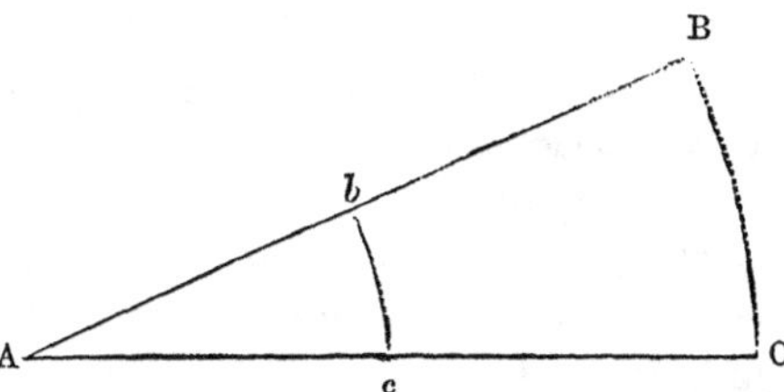

40°; then BC being taken the sine of 90°, *bc* would be the sine of 40°.

And if AB were the radius of a circle, and A*b* the side of an octagon inscribable in the same; then BC being the radius of another circle, *bc* would be the side of an octagon inscribable in the same.

And hence, though the lateral scale AB is fixed, yet a parallel scale BC is obtainable at pleasure.

The manner of taking distances from the sectoral lines will be best understood from the following figure, which contains a portion of the line L.

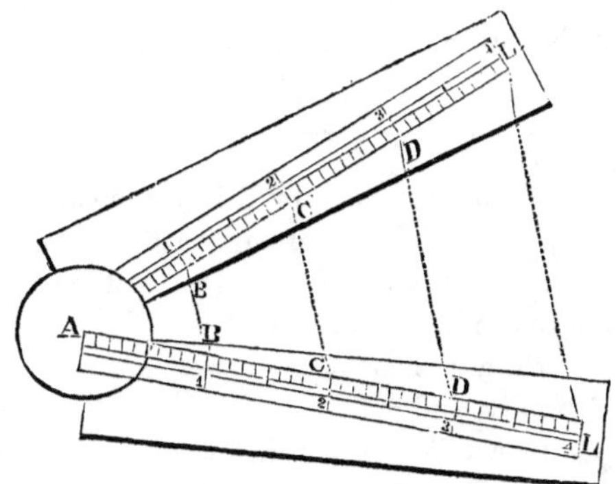

When the distance is taken from the centre A, along

either of the legs to any point, BC, or D, it is called a *lateral distance;* when it is taken from any point of one leg to the corresponding point of the other leg, as from B to B, from C to C, from D to D, or from L to L, it is called a *parallel distance.* To make any length then, as for instance an inch, a parallel distance from 3 to 3, or D to D, the points of the compasses are to be opened an inch; then placing one point against the 3 of one leg of the rule, open the rule till the other point will fall upon the 3 of the other leg. Observe, the points of the compasses must always be placed upon the lines nearest the opening of the rule—upon those which would, if continued, meet in the centre. As distances are taken from all the sectoral lines in the same manner, and as they are of most extensive utility when used conjointly, it will not be necessary to treat of them separately.

USES OF THE SECTOR.

To divide a given line AB into any number of equal parts, as 6.

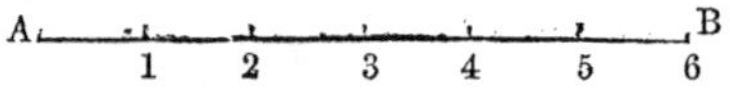

A ———————————————— B
1 2 3 4 5 6

Take the distance from A to B in the compasses, and make it a parallel distance from 6 to 6 on the line L; then the parallel distance from 1 to 1 will be the sixth part of the line AB, as required.

To find a third proportional to two numbers, 4 and 6. Take 6 laterally, and make it a parallel distance from 4 to 4 on L: then take the parallel distance from 6 to 6, and apply it laterally: it will be found to measure to 9, the third proportional required. Or make the lateral distance 4, a parallel distance between 6 and 6; then will the lateral distance 6 be found a parallel distance between 9 and 9.

To find a fourth proportional to three given numbers, 3, 4, and 6. Take the lateral distance 3, and stretch it as a parallel from 4 to 4 on L; then take the lateral distance 6, it will be found to extend as a parallel from 8 to 8. Or make 4 a parallel distance from 3 to 3; then will the parallel from 6 to 6 measure laterally to 8.

The reason is obvious; for in all cases,—

Any lateral distance : its parallel distance : : any other lateral distance : its parallel distance.

And conversely,

Any parallel distance : its lateral distance : : any other parallel distance : its lateral distance.

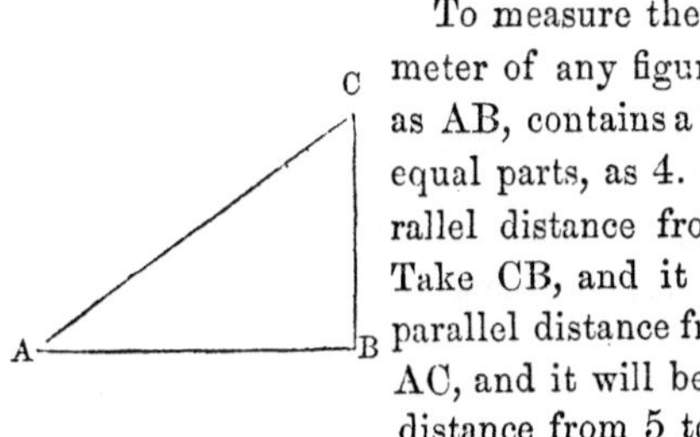

To measure the lines of the perimeter of any figure, one of which, as AB, contains a given number of equal parts, as 4. Make AB a parallel distance from 4 to 4 on L. Take CB, and it will be found a parallel distance from 3 to 3; take AC, and it will be found a parallel distance from 5 to 5.

To multiply any number, 3, by another, 7. Make the lateral distance 3 a parallel distance from 1 to 1 on L; then take the parallel distance from 7 to 7; it will be found the lateral distance of 21.

To divide any number, 40, by another, 5. Make the parallel distance of 40 the lateral distance of 5 on L;

then will the parallel distance of 1 be the lateral distance of 8.

To divide a line in any required proportion, as 2, 3, and 4. Take the length and make it a parallel distance from 9 to 9, their sum, on the line L; then will the parallel distance from 2 to 2, 3 to 3, and 4 to 4, be the parts required.

If the number is greater than 100, take some aliquot part of it, and then multiply the result by the number by which it was divided.

To measure an angle ABC with the line of chords, C. With any radius BA, describe the arc AC; make BA a parallel distance from 60 to 60 on C; then take AC, and moving it along, find the numbers to which it will apply as a parallel distance.

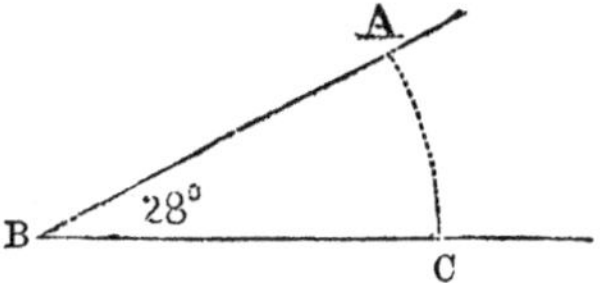

At a given point B, in a given line AB, to make an angle containing any number of degrees, as 30°.

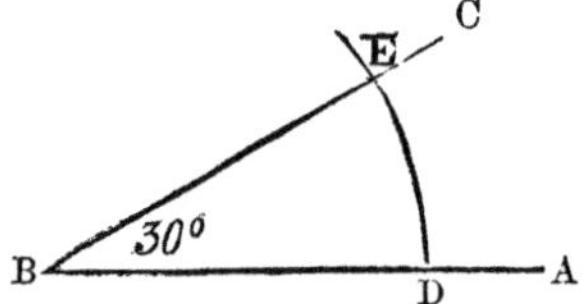

Open the sector to any distance, and take the parallel distance from 60 to 60, with which describe the arc DE. Take the parallel distance from 30 to 30; and setting one foot of the compasses on D, prick off the distance to E; through E draw BC. ABC is the angle required.

And so of all the problems given under the protractor.

To measure any angle, ABC, with the line of sines.

From any point, A, let fall the line AD, perpendicular to BC; then in AD the sine of B, the radius being BA.

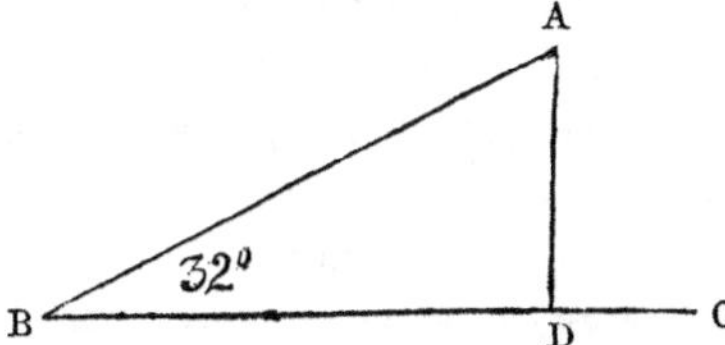

Therefore, take BA, and make it a parallel distance from 90 to 90, on the line S; then take AD, and moving it along, find the numbers to which it will apply as a parallel distance.

At a given point A, in a given line AB, to make an angle containing any number of degrees, as 30.

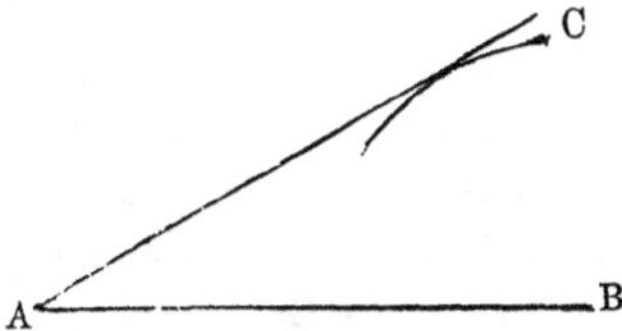

Make AB a parallel distance from 90 to 90 on S; take the parallel distance from 30 to 30, and from B describe an arc. Draw AC touching the arc. CAB is the angle required.

And so of all the problems under the protractor.

6

To measure any angle, ABC, with the line of tangents, T.

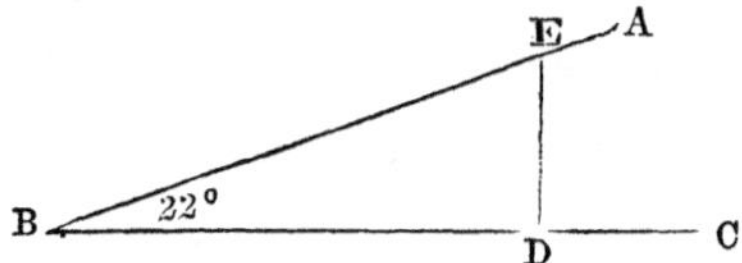

At any distance, D, raise the perpendicular DE, to meet AB; then is DE the tangent of B, the radius being BD. Make BD a parallel distance from 45 to 45 on T; then take DE, and moving it along, find the numbers to which it will apply as a parallel distance.

At a given point A, in a given line AB, to make an angle containing any number of degrees, as 30.

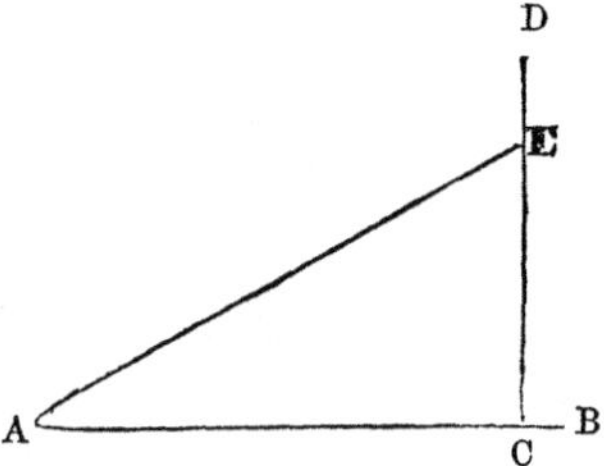

Take any point, C, at which erect the perpendicular CD, Make AC a parallel distance from 45 to 45. Take the parallel distance from 30 to 30, and from C cut off CE equal to it. Join AE. EAB is the angle required.

And so of the rest under the protractor.

To measure any angle, BAC, with the line of secants.

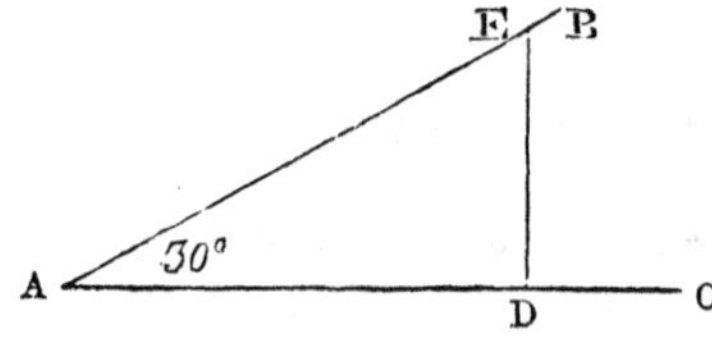

Take any point, D, and raise the perpendicular DE. Make AD a parallel distance from 0 to 0 on the line of secants; then take AE, and moving it along, find the numbers to which it will apply as a parallel distance.

At a given point, A, in a given line AB, to make an angle containing any number of degrees, as 40.

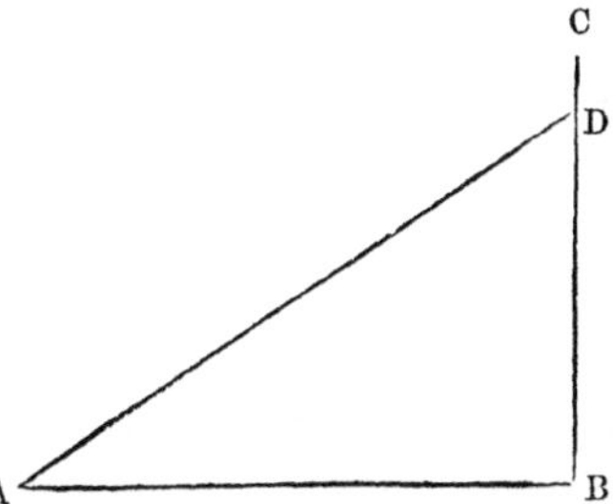

Raise BC at right angles to AB. Make AB a parallel distance from 0 to 0; then take the parallel distance from 40 to 40, and with it, from the point A, cross BC in D. Join AD. DAB will be the angle required.

And so of the rest under the protractor.

The line of secants cannot be employed to much advantage when the number of degrees is under 30, *nor the line of sines when above* 70, *as is evident from an inspection of the rule.*

To find the chord, sine, tangent, and secant of 30 degrees, to a radius of two inches, AB. (See diagram p. 64.)

Take 2 inches in the compasses, and make it a parallel distance from 60 to 60 on the scale of chords; it will also be a parallel distance from 45 to 45 on the tangents; and from 90 to 90 on the sines. Therefore, taking the parallel distance from 30 to 30 on the line C, it will give the chord

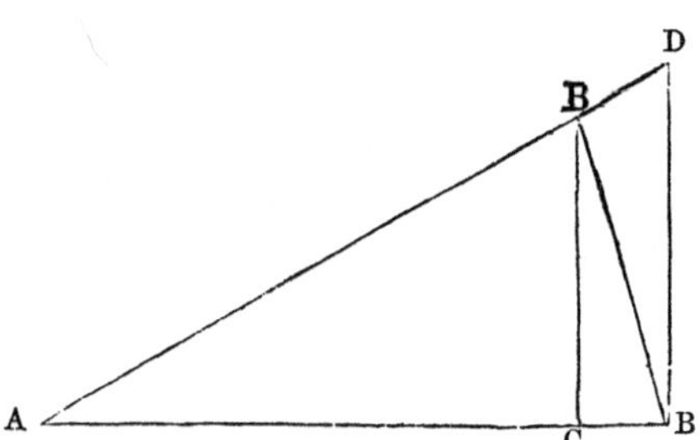

of 30 degrees, BB; from 30 to 30 on the line S, it will be the sine of 30 degrees, BC; from 30 to 30 on the line T, it will be the tangent of 30 degrees, BD. When the number of degrees is above 45, the same opening will not suffice for the tangents; it will then be necessary to set the radius to the 45 of the smaller tangents, and take the aperture as usual. So for the secants, make 2 inches a parallel distance from 0 to 0; then will the parallel distance of 30 be the secant AD.

When the rule is opened to the radius of the smaller tangents, it is also opened to the radius of the secants.

All the problems given under the protractor may be performed by any of these lines; but polygons are most conveniently constructed by the lines Pol.; and in taking distances from these, the points of the compasses are to be placed on the same line as that from which the chords are taken. The side of the hexagon, 6, does not reach to the chord of 60, as has been mentioned; a smaller radius being chosen for the purpose of including the 5 and 4, the pentagon and square. Since the radius of a circle is equal to the side of an inscribed hexagon, the radius is always to be made a parallel distance from 6 to 6.

An example or two will suffice.

To construct a pentagon on a given line AB. Take AB, and make it a parallel distance from 5 to 5. Take the parallel distance from 6 to 6 for a radius, and with it from A and B describe arcs crossing each other in C, and from C describe a circle. The distance AB will run 5 times round it, and form the pentagon required.

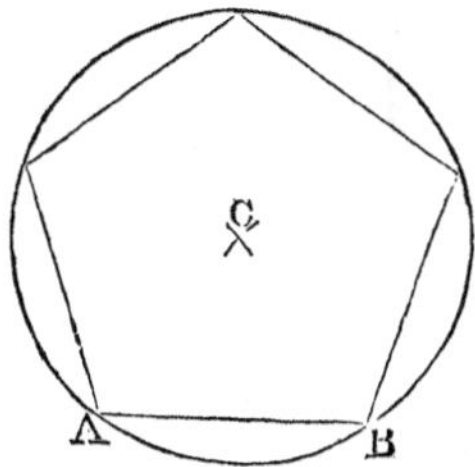

And so of any regular polygon.

In a given circle, to inscribe a regular heptagon. Find the centre of the circle, and make the radius a parallel distance from 6 to 6; take the parallel distance from 7 to 7, and it will run 7 times round the circle as required.

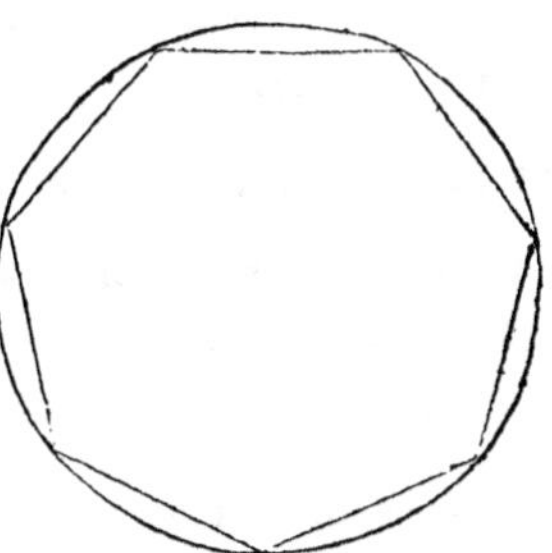

And so of any regular polygon.

APPLICATION OF THE SECTOR TO TRIGONOMETRY.

Required the height of a tower, AB, the angle of elevation, ACB, 200 feet distant, being $47\frac{1}{2}°$.

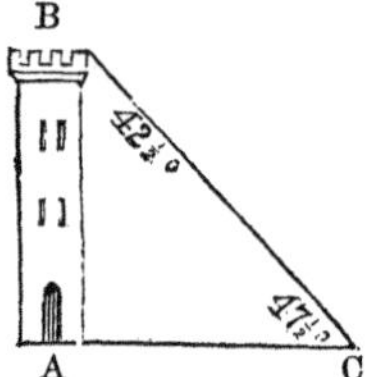

The angle B will be $90 - 47\frac{1}{2} = 42\frac{1}{2}$.
Taking AC radius, AB is the tangent of C, or $47\frac{1}{2}°$.

Hence, Rad. : CA : : tang. $47\frac{1}{2}°$: AB.

Take 200 from a scale of equal parts, (the diagonal scale is the best,) and make it a parallel distance from 45 to 45 on the smaller line of tangents; take the parallel distance from $47\frac{1}{2}$ to $47\frac{1}{2}$, and apply it to the same scale of equal parts; it will be found to measure about 218 feet, the height of the tower.

Otherwise, Sine B : CA : : sine C : AB.

On the line of sines stretch 200 from $42\frac{1}{2}$ to $42\frac{1}{2}$; take the parallel distance from $47\frac{1}{2}$ to $47\frac{1}{2}$, it will measure 218 as before.

What was the perpendicular height of a balloon, when its angles of elevation were 35° and 64°, as taken by two observers at the same time, both on the same side of it, and in the same vertical line, their distance being 880 yards?

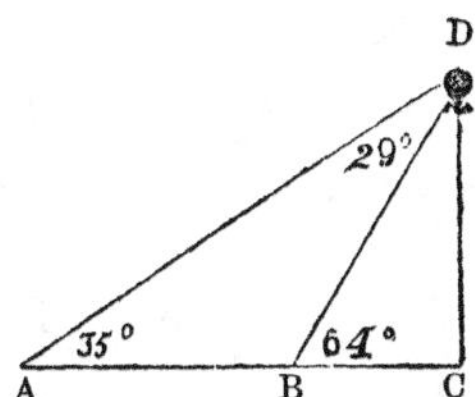

The external angle DBC = DAB + ADB;

∴ ADB = 64 − 35 = 29.

Then, Sine ADB : AB : : sine A : BD.

Take 880 from a scale of equal parts, and make it a parallel distance from 29 to 29 on the line of sines; then will the parallel distance from 35 to 35 measure 1041 for the line BD.

Again, Sine C : BD : : sine DBC : DC.

Take 1041 and make it a parallel distance from 90 to 90, the angle C being a right angle; then take the parallel distance from 64 to 64, and it will measure 936 for CD, the perpendicular height of the balloon.

Wanting to know the distance between two inaccessible trees from the top of a tower, 120 feet high, which lay in the same right line with the two objects, I took the angles formed by the perpendicular wall and lines conceived to be drawn from the top of the tower to the bottom of each tree, and found them to be 33° and 64½°. What is the distance then between the trees?

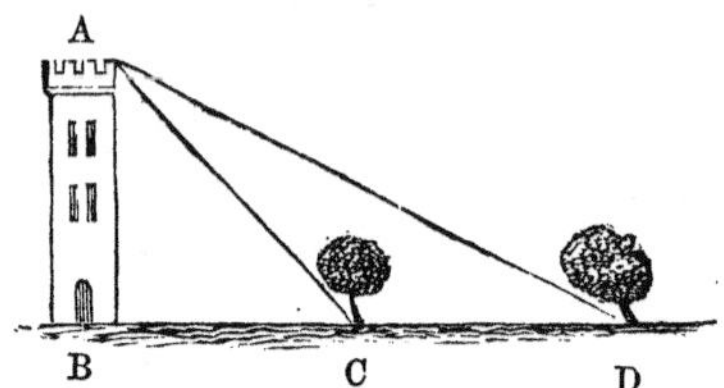

Angle, BAC = 33°. Angle, BAD = 64½.

Making AB radius;—

Rad. : AB : : tang. BAC : BC.

Take 120 from a scale of equal parts, and make it a parallel distance from 45 to 45 on a line of tangents; from which take the parallel distance from 33 to 33; it will measure 78 on the scale for BC, the distance of the first tree from the bottom of the tower.

Again, Rad. : AB : : tang. BAD : BD.

Make AB, or 120, a parallel distance from 45 to 45 on the smaller line of tangents, from which take the parallel distance from 64½ to 64½; it will measure 251½ on the scale for BD, the distance of the farther tree from the bottom of the tower.

Hence 251½ — 78 = 173½, the distance between the trees.

Being on the side of a river, and wanting to know the distance to a house which was seen on the other side, I measured 200 yards in a straight line by the side of the river; and then at each end took the horizontal angle formed between the house and the other end of the line, which were 68° and 73°. What then were the distances from each end to the house?

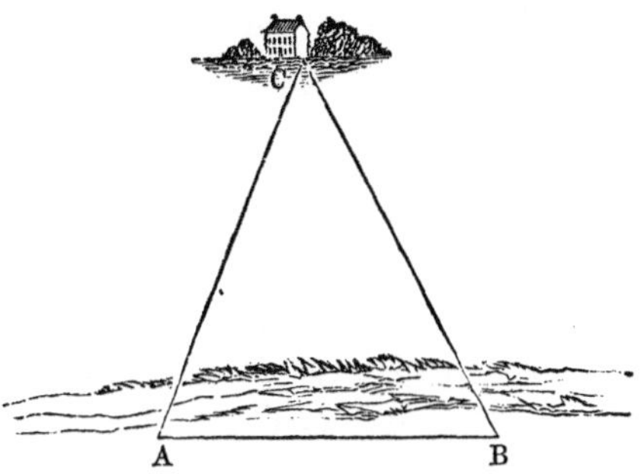

Angle A = 68° ; B = 73° ; 73° and 68° = 141 ;
∴ C = 39°, since the three angles of every triangle = 180°.
Sine C : AB : : sine A : BC ; and
Sine C : AB : : sine B : AC.

Hence, take 200 from the scale, and make it a parallel distance from 39 to 39 on the line of sines ; then will the parallel distance from 68 to 68 measure 294½ equal parts for BC ; and the parallel distance from 73 to 73 will measure 304 for AC.

Having to find the height of an obelisk standing on the top of a declivity, I first measured from its bottom a distance of 40 feet, and there found the angle formed by the oblique plane, and a line imagined to go to the top of the obelisk, 41° ; but, after measuring on in the same direction 60 feet farther, the like angle was only 23° 45′. What was the height of the obelisk ?

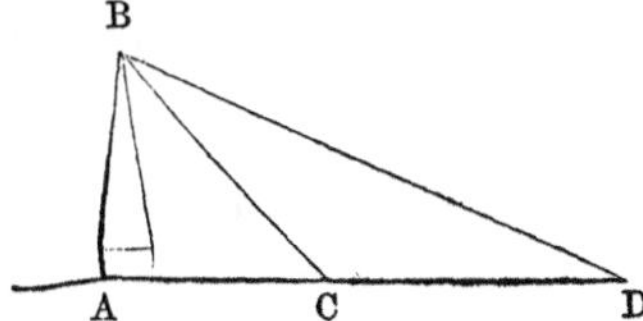

AC = 40; DC = 60; angle BAC = 41°; BDC = 23¾°; ∴ DBC = 17¼°.

Then, Sine DBC : DC : : sine BDC : CB.

Make DC, 60 a parallel distance on the line of sines from 17¼ to 17¼; take the parallel distance from 23¾ to 23¾, and it will measure 81½ for CB.

Then, in the triangle ABC; two sides BC, CA, being known, and their included angle,

Sum of BC, CA : diff. BC, CA : : tang. ½ sum of angles A and B : tang ½ their diff.

Now BC + CA = 121½; and BC — CA = 41½.

Again, since BCA = 41°, the other two = 139°; ½ of which = 69½° for half the sum.

Make 121½ a parallel distance from 69½ to 69½ on the line of tangents; then will 41½ be the parallel distance of 42½. Hence, 69½ — 42½ = 27° for the angle CBA.

Lastly, Sine CBA : CA : : sine BCA : AB.

Make 40 a parallel distance from 27 to 27 on the line of sines; then will the parallel distance from 41 to 41 measure 57¾ for AB, the height of the obelisk.

Standing upon the top of a castle 80 feet high, I threw a string over to a person on the farther side of the moat at the bottom, and found it to measure 100 feet. What was the breadth of the moat, and angle of depression?

Making BC radius; BC : rad. : : BA : sine C.

Make 100 a parallel distance from 90 to 90 on the line

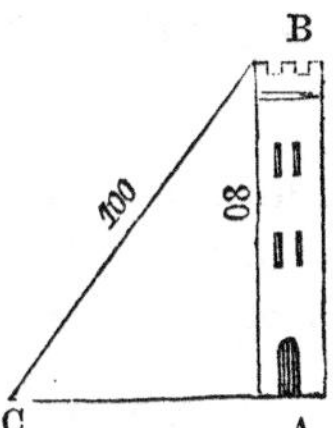

S, then will 80 be a parallel distance from 53 to 53, the angle of depression, or angle C.

Again, Rad. : BC : : cos. C. : AC.

Letting the rule stand, the aperture is set to the first part of the proportion: the cosine of 53° is 37°. Take the parallel distance from 37 to 37, and apply it to the scale: it will measure 60, for CA.

The breadth of a street is 30 feet, the height of a house 40. What must be the length of a ladder that will reach from the top of the house to the opposite side of the way?

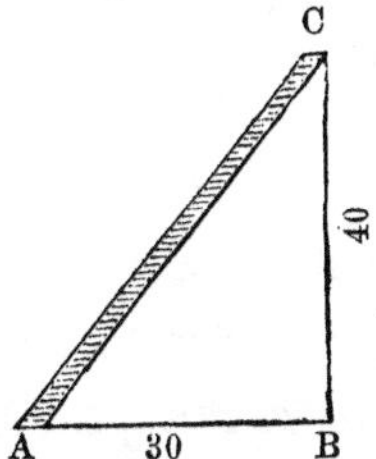

CB + BA = 70; CB — BA = 10.

½ sum of ∠ s, C and A = 45°.

Hence make 70 from the scale a parallel distance from 45 to 45 on the larger line of tangents; then will 10 from the same scale be a parallel distance from 8 to 8.

$$8 + 45 = 53° \text{ the angle CAB.}$$

Then, Rad. : AB : : sec. CAB : AC.

Make 30 a parallel distance from 0 to 0 on the line of secants; then will the parallel distance from 53 to 53 measure 50 feet, the length of the ladder AC.

Coasting along the shore, I saw a cape bear from me *NNE;* then I stood away *N W b W* 20 miles, and I observed the same cape to bear from me *N E b E:* required the distance of the ship from the cape at her last station.

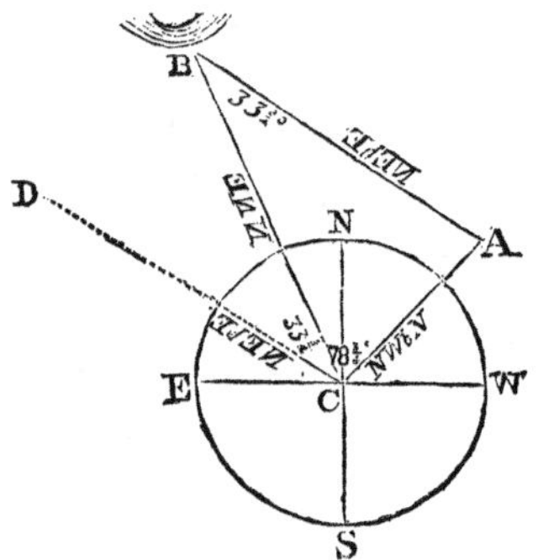

C, the ship's first station; A, the place of the ship in her second station; B, the cape. Then,

CA = 20 miles; ∠ ACB = 78¾; ∠ ABC = ∠ BCD = 33¾.

Hence, Sine ABC : AC : : sine ACB : AB.

Make 20 a parallel distance from 33¾ to 33¾ on the line of sines; then will the parallel distance of 78¾ measure 35 miles nearly, for AB, the distance of the ship from the cape in her last station.

Enough has been shown to enable the pupil to apply the rule to most of the purposes to which it is adapted. Beside the sectoral lines, which may now be dismissed, there are on the edges the decimal parts of a foot, and along the margin a scale of inches; these need no explanation. There are also three other lines, marked N, S, and T, called Gunter's lines. The distances on N are the logarithms of numbers; on S, logarithmic sines; and on T, logarithmic tangents. These are the same as the lines of the slide-rule.

CONSTRUCTION.

Referring to a table of logarithmic numbers, sines, and tangents, we have

No.	Log.	°	Sine.	Tang.
1	.000	1	8.241	8.241
2	.301	2	8.542	8.543
3	.477	3	8.718	8.719
4	.602	4	8.843	8.844
5	.698	5	8.940	8.941
6	.778	6	9.019	9.021
7	.845	&c.	&c.	&c.

Lay down three parallel lines, N, S, T, as below, and draw AB perpendicular to them, that the beginnings of

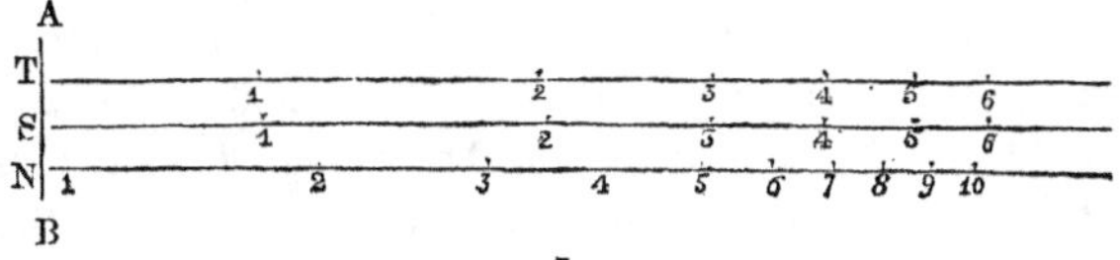

the three may correspond. The log. of 1 being 0, set 1 at the commencement N. From the diagonal scale take off 301, the log. of 2, and lay it from N to 2; turn it over again, and it will mark the log. of 4; turning it again, it will reach to the log. of 8, and so on.

Again, from the diagonal scale take off 477, the log. of 3, and lay it from N to 3; turn it over, and it will reach to 9, from 9 to 27, and so on. Setting the same distance from 2, it will reach to 6, from 6 to 18, and so on. For 5 take 698 from the scale, and set it from N to 5; the same distance will reach from 2 to 10, &c. For 7 take 845, and set it off from N to 7. Lay the line over again, and proceed to fill up the distances 11, 12, &c. from a set of tables, till the line is finished.

For the line S.

The logarithmic sine of 1° is 8.241. Disregarding the whole number 8, which is prefixed to indicate the great extent into which the radius is supposed to be divided, take from the same scale 241, and lay it from S to 1; lay off 542 from S to 2; 718 from S to 3; 843 from S to 4; 940 from S to 5; 1019 from S to 6; and so on to 90°.

For the line T.

The logarithmic tangent of 1° is 8.241; hence, it will be over the 1 of S; lay off 543 from T to 2; 719 from T to 3; 844 from T to 4; 941 from T to 5; 1021 from T to 6; and so on to 45. Arriving at 45, the numbers return, and 50 is placed with 40; 30 with 60; 20 with 70; 80 with 10; and 90 at the beginning; the decimal part of the logarithmic tangent of any degree being the arithmetical complement of the cotangent; since, as was shown, the radius is a mean between the tangent and cotangent.

USES.

An example or two of the manner of using these lines will be sufficient. For the reasons of the operations, the student is referred to the Treatise on the Slide-rule.

In the lines S and T, the numbers show the values which are to be taken. On the line N, the second division will be ten times those of the first division; the values are, otherwise, arbitrary. Thus, if 3 of the first division be 3, that of the second will be 30; if 30, 300; and so on.

Uses of the line N, *or Logarithmic Numbers.*

To multiply 4 by 5. Take the distance from 1 to 4, stretching the rule open to its full extent; it will then reach from 5 to 20, the product.

To divide 30 by 5. Take the distance from 1 to 5, it will reach backwards from 30 to 6, the quotient.

To find a fourth proportional to three numbers, 3, 4, and 6. That is, 3 : 4 : : 6 : ? Take the distance from the first to the third; it will reach from the second to the fourth. The distance from 3 to 6 will reach from 4 to 8, the fourth proportional required.

To square and cube 3. Take the distance from 1 to 3, and set it forwards from 3; it will reach to 9, the square required. Set it forwards again from 9, and it will reach to 27, the cube.

To extract the square root and cube root of 64. Take the distance from 1 to 64; find, by the line of lines, the half of this; it will reach from 1 to 8, the square root.

Find by the line of lines the third of it, and it will reach from 1 to 4, the cube root of 64.

Uses of the lines N, S, *and* T, *conjointly.*

In the right-angled triangle ABC, suppose AB 124, and the angle A 34° 20′; consequently, the angle C 55° 40′. Required BC and CA.

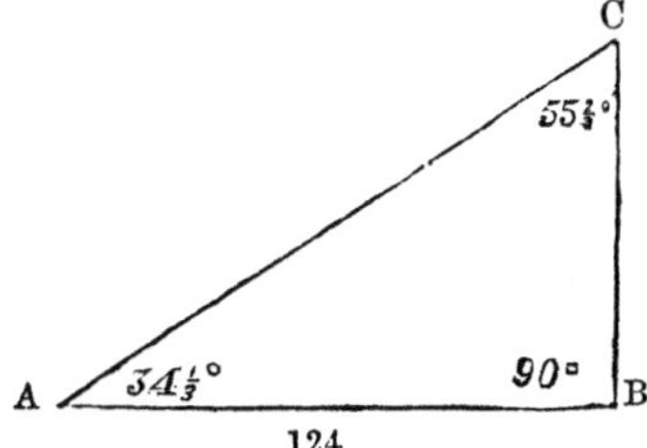

Sine C : sine A : : AB : BC.

Hence, on the line S, take the distance backwards from $55\frac{2}{3}$ to $34\frac{1}{3}$; this will reach back on the line N from 124 to 84.7 the length of BC.

Again, Sine C : sine B : : AB : AC.

Take the distance forwards on the line S from $55\frac{2}{3}$ to 90; this will reach forwards on the line N from 124 to 150, the length of AC.

Otherwise for the perpendicular BC,

Rad. : tang. A : : AB : BC.

On the line T, take the backward distance from 45 to $34\frac{1}{3}$; this will reach back, on the line N, from 124 to 84.7 for BC, as before.

From the top of a tower, by the sea-side, of 143 feet

height, it was observed that the angle of depression of a ship's bottom, then at anchor, measured 35° : what then was the ship's distance from the bottom of the wall?

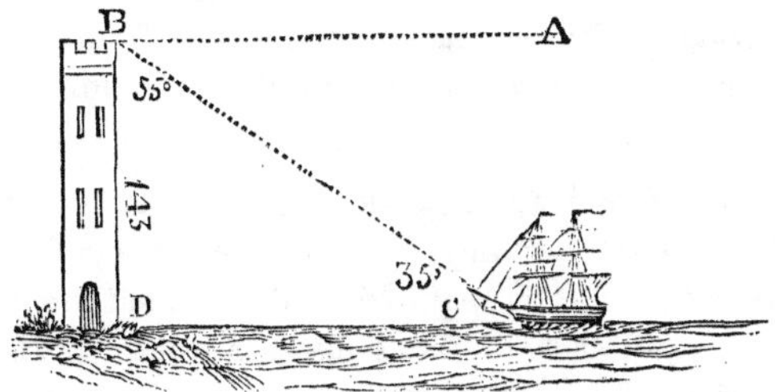

The angle of depression of the vessel is ABC, and consequently is equal to the angle of elevation of the tower, BCD. Hence, making BD radius;

Rad. : tang. 55° : : BD : DC.

Stretch the compasses on the line T, from 45 to 55; this will reach from 143 to 204 on the line N.

What is the perpendicular height of a hill; its angle of elevation, taken at the bottom of it, being 43°; and 200 yards farther off, on a level with the bottom of it, 31°?

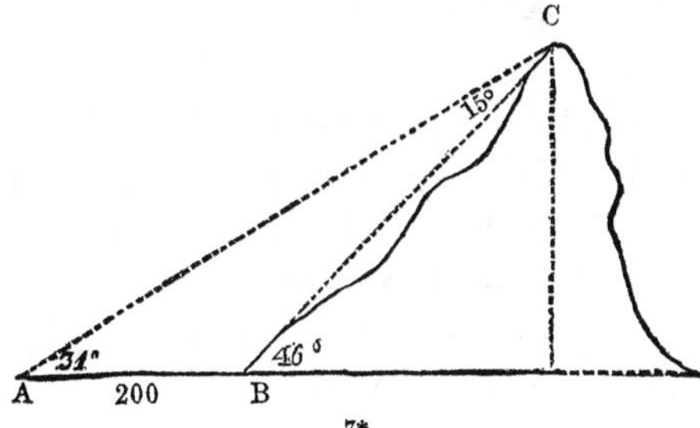

Sine 15° : sine 31° : : 200 : BC; and
Sine 90 : sine 46° : : BC : CD.

Stretch the compasses on the line S from 15 to 31; this distance, on the line N, will reach from 200 to 398 for BC. Again, take the backward distance from 90 to 46 on S; this will reach back from 398 to 286 for CD, the height of the hill.

A point of land was observed, by a ship at sea, to bear *E b S;* and after sailing *N E* 12 miles, it was found to bear *S E b E.* It is required to determine the place of that headland, and the ship's distance from it at the last observation.

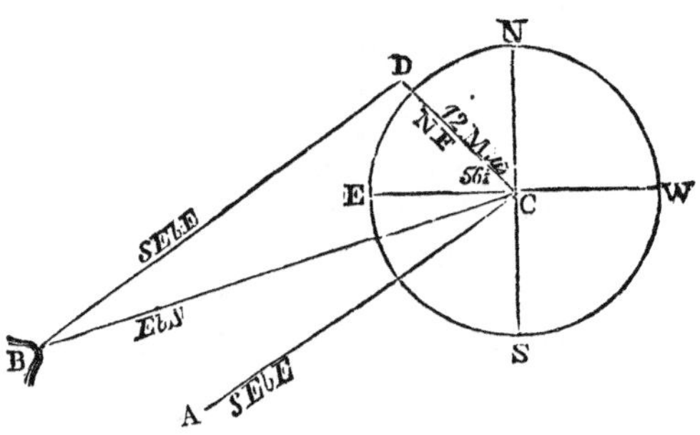

Sine 22½ : sine 56¼ : : 12 : BD.

On the line S take the forward distance from 22½ to 56¼; this will reach from 12 to 26 for BD, the ship's distance from the last place of observation.

Standing upon the top of a castle 80 feet high, I threw a string over to a person on the farther side of the moat at bottom, and found it to measure 100 feet. Required the breadth of the moat, and the angle of depression.

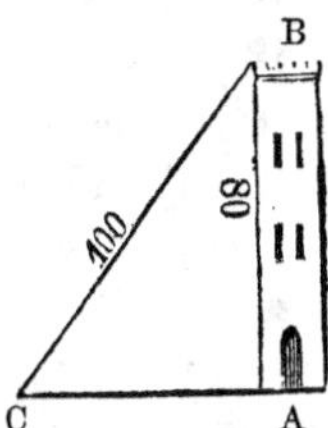

Take CB radius; then—

CB : BA : : rad. : sine C; and
Rad. : cos. C : : CB : CA.

Measure back on N from 100 to 80; this on S will reach back from 90 to 53 for the sine. The cosine of 53 to 37.

On S take the distance back from 90 to 37; it will reach back on N, from 100 to 60 for CA.

These examples will be sufficient to show the pupil the method of using the lines. He can work over the questions that were solved by the natural sines and tangents of the sectoral lines; taking the numbers from N, instead of from a scale of equal parts, and the sines and tangents from S and T respectively; observing, that a *forward* distance from one is to be applied as a *forward* distance to the other, and a *backward* distance from one as a *backward* distance to the other. The distance on the line T, from 45 to 55, is to be reckoned as a forward distance, while the distance from 45 to 35, though the same, is to be accounted a backward distance. He will also observe, in stating the proportion, that that which was made the second term for the sectoral lines, must be placed third when intended for the Gunter's. Thus—

For the sectoral, AB : sine C : : BC : sine A.
For the logarithmic, AB : BC : : sine C : sine A.

Such is the Sector; it was devised by the celebrated Gunter, about the year 1607, and did not, at first, contain the N, S, and T lines; these were added sixteen years later, or nine years after Napier's admirable invention of logarithms. In the year 1657, Partridge greatly improved upon the plan, by laying the lines down double, and sliding one against the other. The sector dwindles into insignificance, in comparison with the slide-rule, which is nearly a perfect instrument, and adapted, in a degree, to every species of computation for which logarithms are available. The chief merit, however, is due to Gunter, for hundreds can improve where only one can invent.

LOGARITHMS.

LOGARITHMS are a series of numbers in arithmetical progression, corresponding to another series of numbers in geometrical progression: thus—

0	1	2	3	4	5	6	7	8
1	2	4	8	16	32	64	128	256

where the indices 0, 1, 2, 3, &c. in the arithmetical series are the logarithms of the numbers 1, 2, 4, 8, &c. of the geometrical.

On examining these, it will be found that if any two of the logarithms, or indices, are added together, their *sum* will be the logarithm or index of the *product* of the numbers to which they belong. Thus, 2 and 3 are 5; the number against this is 32, which is the product of 4 and 8, the numbers beneath the indices 2 and 3.

In like manner, if any one of the indices be *subtracted* from another, their difference is the index of the *quotient* of the numbers. Thus, 5 from 7 leave 2, the number against which is 4, the quotient of 128 by 32.

For the same reason, if any one of the indices be *multiplied* by another denoting *power*, the product will be the index of that power. Thus, to find the square of 8; its index is 3, which, doubled, becomes 6, the index of 64, the square of 8, as required.

Lastly, if the index of any number be *divided* by the index of any *root*, the quotient will be the index of that root. Thus, to find the square root of 16; its index is 4, the half of which is 2, which is the index of 4, the square root of 16, as required. From which it appears that addition of logarithms answers to multiplication of common numbers, subtraction to division, multiplication to involution, and division to evolution. The same will also be the case if we select any other geometrical series; thus—

$$\overset{0}{1}\quad \overset{1}{3}\quad \overset{2}{9}\quad \overset{3}{27}\quad \overset{4}{81}\quad \overset{5}{243}\quad \overset{6}{729}$$

or—

$$\overset{0}{1}\quad \overset{1}{10}\quad \overset{2}{100}\quad \overset{3}{1{,}000}\quad \overset{4}{10{,}000}\quad \overset{5}{100{,}000}\quad \overset{6}{1{,}000{,}000}$$

From which it is evident that the same indices may serve for any system, and, consequently, that the varieties of systems of logarithms are infinite.

Now, since numbers are but expressions for the measure of distances from a fixed point, it follows, that if from the commencement 0 of any line, we set off equal distances in the points 1, 2, 3, &c., and raise against them a series of perpendiculars, 1, 2, 4, 8, &c., we shall have in the extremities of these perpendiculars a series of distances 1, 2, 4, 8, &c., whose logarithms will be the distances 0, 1, 2, 3, &c. These representing the indices of the distances measured by the perpendiculars, they will of course possess the same properties as the indices themselves.

Thus, let it be required to multiply 16 by 4. With a pair of compasses take the distance from 0 to 2, the index corresponding to 4; set one foot of the compasses on 4, the index of 16, the other point will reach *forwards* to 6, the index of 64, the product of the numbers 4 and 16.

Again, let it be required to divide 64 by 16. Take the distance from 0 to 4, the index of the less; place one foot

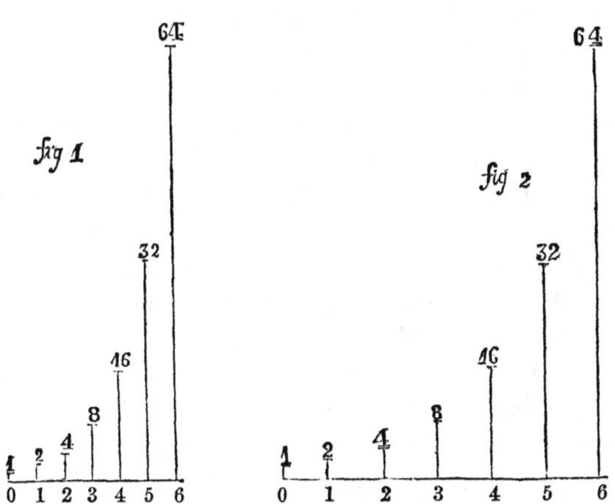

of the compasses on 6, the index of the greater, the other point will reach *back* to 2, the index of 4, the quotient required. Next, let it be required to find the square of 8. Take the distance from 0 to 3, the index of 8. Place the compasses on the point 0, and turn them over twice; they will reach to 6, the index of 64, the square required.

Lastly, to find the square root of 64. Take half the distance from 0 to 6, which is the point 3, the index of 8, the square root required.

But for taking squares, and square roots, it will be more convenient to have the logarithmic distances laid down to two scales, *Figs.* 1 and 2, in which the distances of the latter shall be twice those of the former. Then to take the square of 8. Measure the distance from 0 to 3, *Fig.* 2; apply this to *Fig.* 1; and it will reach to 6, the index of 64, the square required. And to find the square root of 64. Take the distance from 0 to 6, *Fig.* 1; it will reach from 0 to 3, *Fig.* 2, which is the index of 8, the square root required. Thus, by mechanical means, we obtain the same results as by arithmetical calculation: a forward motion performs the work of multiplication; a backward, that of subtraction; an increase of distance, the raising of powers; and a diminution, the extraction of roots.

But since the application of the compasses is a tedious method, it is desirable to perform the same operations by a readier way, which we may now proceed to consider.

If to the end of AB we set the beginning of CD, it is evident that the distance AE, reaching from the commencement of one to the end of the other, measures their united lengths, or expresses the sum of the two;

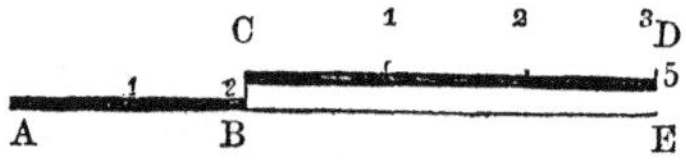

C 1 2 3 D
1 2 5
A B E

That is, AB + CD = AE = 2 + 3 = 5.

And if against E, the extremity of the line AE, we set D, the extremity of CD; the part AB, between the beginning of the longer and the beginning of the shorter, measures their difference;

That is, AE — CD = AB = 5 — 3 = 2.

For addition, then, the rule may be expressed:—Set the *beginning* of one to the *end* of the other; then against the *end* of the *second* is their *sum* on the first. And, for subtraction:—Set the *ends together*, then at the *beginning* of the *shorter* is their *difference* on the *longer*.

A few operations will best imprint this upon the memory of the student, for which purpose he must furnish himself with scales of equal parts, as on p. 85.

On a sheet of paper take any distance, and set it off 20 times, as A. And beneath it at any convenient distance,

Common Scale.

1 2 3 4 5 6 7 8 9 10 11 12 13 14 15 16 17 18 19 20 A

1 2 3 4 5 6 7 8 9 10 D

1 2 3 4 5 6 7 8 9 10 11 12 13 14 15 16 17 18 19 20 B

Slide.

double the space 10 times, as D. Then take a slip of card or paper, B, which we may call the *slide*, of a breadth about sufficient to fill up the space left between A and D, and mark upon it the same scale of equal parts as upon A. With these we may perform the following operations:—

I. To find the half of any number, and the double of any number.

Lay the slide in the space between A and D, then under any number on B is its half on D, and over any number on D is its double on B.

II. To find the sum of any two numbers, and half their sum.

Let the numbers be 6 and 4. Against the 6 on A set the beginning on B, then over the 4 on B is 10, their sum, and under it on D is 5, half their sum.

III. To find the difference of any two numbers, and half their difference.

Let the numbers be 11 and 5. Against the 11 on A set the 5 on B, then over the beginning of B is 6, their difference, and under it 3, half their difference.

IV. The sum of two numbers may be found by three methods.

Thus, of 5 and 7. Against 5 on A set the beginning of B, over the 7 on B is 12 on A;—or push the slide out till the 5 on B reach the beginning of A, then under the 7 on A is 12 on B;—or invert the slide, and set the 7 on B to the 5 on A, then against the beginning of B is 12 on A. Also for the difference; under the 7 on A set the 5 on B, against the beginning of B is 2 on A;—or against the 5 on A set the 7 of B, against the beginning of A is 2 on B;—or invert the slide and set its beginning to the 7 of A, then over the 5 of B is 2 on A.

V. To add any number, 3, to twice another, 4.

Push the slide back till the 3 on B reaches the beginning of D; over the 4 on D is 11 on B. Or against the 4 on D set the beginning of the slide: over the 3 of this is 11 on A.—Hence, to multiply any number by 3, as 6; (that is, to add 6 to twice 6.) To the 6 on D set the 6 on the slide inverted; over the beginning of this is 18 on A.—Hence also to divide any number by 3, as 18. Set the beginning of the slide inverted to 18 on A, or the 18 on B to the beginning of A or D, then where equal values meet together on B and D, is the third of the number; thus, the 6 on B meets the 6 on D.

VI. To add a number, 5, to the half of another, 6.

Against 5 on D set the beginning of B, under the 6 of which is 8 on D.

VII. To subtract a number, 3, from twice another, 4.

Against the 4 on D set the 3 of B, over the beginning of which is 5 on A.

VIII. To subtract half a number, 6, from another, 7.

Against the 7 on D set the 6 of B; under the beginning of this is 4 on D.

IX. From any number, 9, to subtract twice another, 2.

Against 9 on B set the 2 of D; over the beginning of D is 5 on B: or invert the slide, and over the beginning of D set 9 on B; over the 2 of D is 5 on B.

X. To subtract a number, 3, from the sum of two others, 4 and 5.

Against the 3 on A set the 4 on B; then under the 5 of A is 6 on B. The reason of this is plain; for to have added 4 to 5, the slide ought to have been pushed out till the 4 fell under the beginning of A; but, as it was not removed so far back by 3 spaces, the result will evidently be 3 less than the sum. Otherwise, invert the slide, and against the 4 on A set the 5 of B; over the 3 of this is 6 on A, or under the 3 of A is 6 on B. Here, had we wanted the sum, we should have counted to the beginning of the slide, or from the beginning of A; but as in both instances we omitted 3 spaces, the result is 3 less than the sum.

XI. To subtract 3 from 5 added to twice 4.

Under the 3 of A set the 5 of B; over the 4 of D is 10 on B. The reason of this is obvious: to have added 5 to twice 4, the slide ought to have been pushed out till the 5 reached the beginning of A or D, but as it was not removed so far by 3 spaces, the result is 3 less than the sum. It may be performed otherwise by inverting the slide. Against the 4 of D set the 5 of B; over the 3 of this is 10 on A, or under the 3 of A is 10 on B. The reason is evident.

XII. To subtract twice a number, 3, from the sum of 5 and twice 4.

Place the 5 of B over the 3 of D; over the 4 of D is 7 on B. To have added 5 to twice 4, the 5 of B ought to have been set at the beginning of D, but as it is not removed so far back by twice 3 spaces, the result is twice 3 less than the sum.—These operations are to be carefully attended to, especially the last, as, in working with the slide-rule, it is more employed than any other.

XIII. To substract half 6 from 5 added to half 8.

Against the 5 on D set the 6 of B; under the 8 of B is 6 on D. To have added 5 to half 8, the beginning of the slide ought to have been placed at the 5 of D, then under the 8 of B would have been the sum; but as the slide is not set so forward by 6 half-spaces, the result is half 6 less than the sum.

As the whole art of using the slide-rule depends upon a perfect knowledge of these simple movements, the pupil will do well to make himself thoroughly acquainted with them, and to attend carefully to the *reasons* for every operation. Unless he does this, he must always work in the dark, and will be perpetually liable to fall into mistakes; whereas, if he makes himself intimate with them, he will be enabled to proceed with certainty and pleasure.

We may now resume the consideration or logarithms,

and return to FIGS. 1 and 2, on page 83; and here it is at once evident, that as far as the solution of any questions is concerned, the perpendiculars are of no importance, the *equal* distances and the *increasing* numbers being all that are required. If, then, to the scales we have been using, or to any others of equal parts, where the distances on D are double those of A and B, we affix the geometrical numbers 1, 2, 4, 8, &c., the distances, measured from the commencement, will be the logarithms of those numbers, and may be applied to the usual purposes for which logarithms are adapted.

The preceding operations may now be repeated; but, instead of simply adding and subtracting, doubling and halving, they will present themselves under the shapes of multiplying and dividing, squaring and extracting the

Logarithmic Scale.

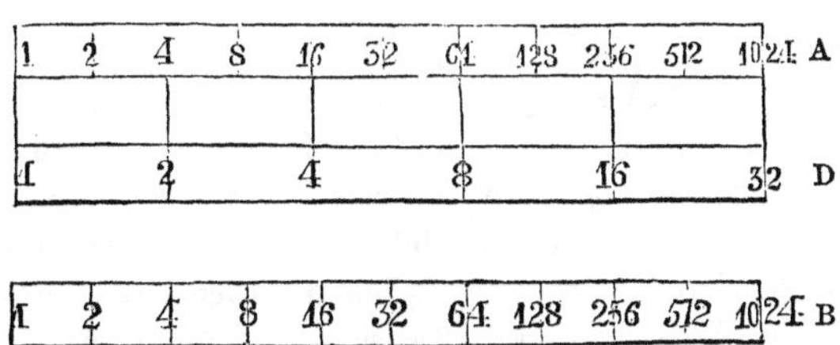

Slide.

square root; and will become converted into the following operations, which the pupil should compare with the preceding, step by step, as he advances, making use of the logarithmic scale and slide, and the common scale and slide, alternately.

I. To find the square of any number, and the square root of any number.

Lay the slide in the space between A and D; then under any number on B is its square root on D; and over any number on D is its square on B.

II. To find the product of any two numbers, 4 and 16, and the square root of their product.

To 4 on A set the beginning of B, over the 16 of which is 64, their product on A; and under it, 8, the square root of their product, the mean proportional between them.

III. To find the quotient of any two numbers, 8 and 128, and the square root of their quotient.

Against the 128 on A set the 8 on B; then over the beginning of B is 16, their quotient; and under it, 4, the square root of their quotient.

IV. The product of two numbers may be found by three methods.

Thus, of 16 and 4. Against 16 on A set the beginning of B; over the 4 of B is 64 on A;—or, push the slide out till the 4 on B reaches the beginning of A; then under the 16 of A is 64 on B;—or, invert the slide, and set the 4 on B to the 16 on A; then against the beginning of B is 64 on A. Also, for the quotient:—under the 16 on A set the 4 on B; against the beginning of B is 4 on A;—or, against the 4 on A set the 16 of B; against the beginning of A is 4 on B;—or invert the slide and set its beginning to the 16 of A; then over the 4 of B is 4 on A.

V. To multiply any number, 4, by the square of another, 8.

Push the slide back till the 4 on B reaches the beginning of D; over the 8 on D is 256 on B; or against the 8 on D set the beginning of the slide; over the 4 of this is 256 on A.

Hence, to cube any number, as 8, that is, to multiply 8 by the square of 8. Against the beginning of D set 8 on B; over the 8 of D is 512 on B;—or, invert the slide, and against 8 on D set 8 on B; over the beginning of B is 512 on A.

Hence, to find the cube root of a number, as 512. Set the beginning of the slide inverted to 512 on A, or 512 on B to the beginning of A or D; then where equal values meet together on B and D is the cube root of the number, 8.

VI. To multiply a number, 8, by the square root of another, 16.

Against the 8 on D set the beginning of B, under the 16 of which is 32 on D.

VII. To divide the square of any number, 16, by any number, 8.

Against the 16 on D set the 8 of B, over the beginning of which is 32 on A.

VIII. To divide any number, 64, by the square root of any number, 16.

Against the 64 on D set the 16 of B; under the beginning of this is 16 on D.

IX. To divide any number, 64, by the square of another, 4.

Against 64 on B set the 4 of D; over the beginning of D is 4 on B.

X. To find a fourth proportional to three numbers, 4, 32, and 64; or, which is the same thing, to divide by 4, 32 times 64.

Against the 4 on A set the 32 on B; then under the 64 on A is 512 on B. For inverse proportion,—as, if 8 men can build a wall in 16 days, in how many days will 32 perform the same?—Invert the slide, and against the 8 on A set the 16 of B; under the 32 of this is 4 on B; or, over the 32 of B is 4 on A.

XI. To divide by 8, 32 times the square of 4.

Under the 8 of A set the 32 of B; over the 4 of D is 64 on B;—or, invert the slide, and against the 4 of D set the 32 of B; over the 8 of this is 64 on A, or under the 8 of A is 64 on B.

XII. To divide by the square of 4, 8 times the square of 16.

Place the 8 of B over the 4 of D; over the 16 of D is 128 on B.

XIII. To divide by the square-root of 64, 4 times the square root of 256.

Against the 4 of D set the 64 of B; under the 256 of B is 8 on D.

These operations, it will be seen at once, are precisely the same as the former. They may be represented algebraically, as beneath; where m, n, and r, are any numbers taken at pleasure.

I. $$\text{Log. } m^2 = 2 \text{ log. } m;$$
$$\therefore m \text{ on D, } m^2 \text{ on B.}$$
$$\text{Log. } \sqrt{m} = \tfrac{1}{2} \text{ log. } m;$$
$$\therefore m \text{ on B, } \sqrt{m} \text{ on D.}$$

II. $\text{Log. } m n = \text{log. } m + \text{log. } n;$

$\therefore m$ on A + n on B, $m n$ on A.

$$\text{Log. } \sqrt{m n} = \frac{\text{log. } m + \text{log. } n}{2};$$

$\therefore m$ on A + n on B, $\sqrt{m n}$ on D.

III. $\text{Log. } \frac{m}{n} = \text{log. } m - \text{log. } n;$

$\therefore m$ on A — n on B, $\frac{m}{n}$ on A.

$$\text{Log.} \sqrt{\frac{m}{n}} = \frac{\text{log. } m. - \text{log. } n;}{2}$$

$\therefore m$ on A — n on B, $\sqrt{\frac{m}{n}}$ on D.

IV. $\text{Log. } m n = \text{log. } m + \text{log. } n;$

$\therefore m$ on A + n on B, $m n$ on A.

Or, m on B + n on A, $m n$ on B,

and $\text{log. } \frac{m}{n} = \text{log. m.} - \text{log. } n;$

$\therefore m$ on A — n on B, $\frac{m}{n}$ on A.

Or, m on B = n on A, $\frac{m}{n}$ on B.

V. $\text{Log. } m n^2 = \text{log. } m + 2 \text{ log. } n;$

$\therefore m$ on B + n on D, $m n^2$ on B.

Or, n on D $+$ m on B, $m\,n^2$ on A,

and log. m^3 or $m\,m^2 =$ log. m $+$ 2 log. m;

$\therefore$ m on B $+$ m on D, m^3 on B.

Or, m on D $+$ m on B, m^3 on A.

VI. $$\text{Log. } m\,n^{\frac{1}{2}} = \text{log. } m + \frac{\text{log. } n}{2};$$

$\therefore$ m on D $+$ n on B, $m\,n^{\frac{1}{2}}$ on D.

VII. $$\text{Log. } \frac{m^2}{n} = 2 \text{ log. } m - \text{log. } n;$$

$\therefore$ m on D $-$ n on B, $\frac{m^2}{n}$ on A.

VIII. $$\text{Log. } \frac{m}{n^{\frac{1}{2}}} = \text{log. } m - \frac{\text{log. } n}{2};$$

$\therefore$ m on D $-$ n on B, $\frac{m}{n^{\frac{1}{2}}}$ on D.

IX. $$\text{Log. } \frac{m}{n^2} = \text{log. } m - 2 \text{ log. } n;$$

$\therefore$ m on B $-$ n on D, $\frac{m}{n^2}$ on B.

X. $$\text{Log. } \frac{m\,n}{r} = \text{log. } m + \text{log. } n - \text{log. } r;$$

$\therefore$ $-\,r$ on A $+$ m on B $+$ n on A, $\frac{m\,n}{r}$ on B.

XI. $$\text{Log. } \frac{m\,n^2}{r} = \text{log. } m + 2 \text{ log. } n - \text{log. } r;$$

$\therefore$ $-\,r$ on A $+$ m on B $+$ n on D, $\frac{m\,n^2}{r}$ on B.

XII. $\text{Log.} \frac{m\,n^2}{r^2} = \log. m + 2 \log. n - 2 \log. r;$

$$\therefore -r \text{ on D} + m \text{ on B} + n \text{ on D}, \frac{m\,n^2}{r^2} \text{ on B.}$$

XIII. $\text{Log.} \frac{m\sqrt{n}}{\sqrt{r}} = \log. m + \frac{\log. n}{2} - \frac{\log. r}{2};$

$$\therefore -r \text{ on B} + m \text{ on D} + n \text{ on B}, \frac{m\sqrt{n}}{\sqrt{r}} \text{ on D.}$$

In all these operations the student will at once perceive, what it is scarcely necessary to mention, that the movements are from the paper to the slide, and from the slide to the paper, alternately.

Thus, from A to B, and back to A. From A to B, and thence to D. From B to D, and back to B. From D to B, and back to D. From D to B, and thence to A. From A to B, and thence to A, and back to B. From A to B, and thence to D, and back to B.

The preceding operations include the whole theory of the Slide-rule; but as it is suitable only for particular numbers, in the form we have presented it, it remains to explain the method of inserting those that have been omitted. For this purpose, draw any angle BAC, and in

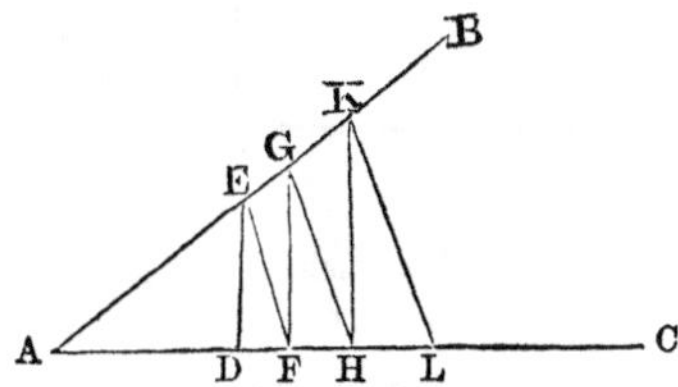

the base take any two points, D, F; make AE = AF; join DE, EF, and through F draw FG parallel to DE;

and through G draw GH parallel to FE; and so on. Then we have,

$$AD : AE :: AF : AG. \quad \text{But, } AE = AF;$$
$$\therefore AD : AF :: AF : AG. \quad \text{But, } AG = AH;$$
$$\therefore AD : AF :: AF : AH.$$

Hence, AD : AF : : AF : AH : : AH : AL, &c.

Putting AD = 1, and AF = a, we have

$$1 : a :: a : a^2 :: a^2 : a^3 :: a^3 : a^4, \text{ \&c.};$$

or, since $a^0 = 1$,

$$a^0 : a^1 :: a^1 : a^2 :: a^2 : a^3 :: a^3 : a^4, \text{ \&c.};$$

where the indices, 0, 1, 2, 3, 4, &c. are the logarithms of a^0, a^1, a^2, a^3, a^4, &c., or of AD, AF, AH, AL, &c.

Along any line, MN, from any point 0, set off a number of equal distances, 1, 2, 3, 4, &c., and at these erect perpendiculars, taking the first equal to AD; the next, AF; the next, AH; and so on.

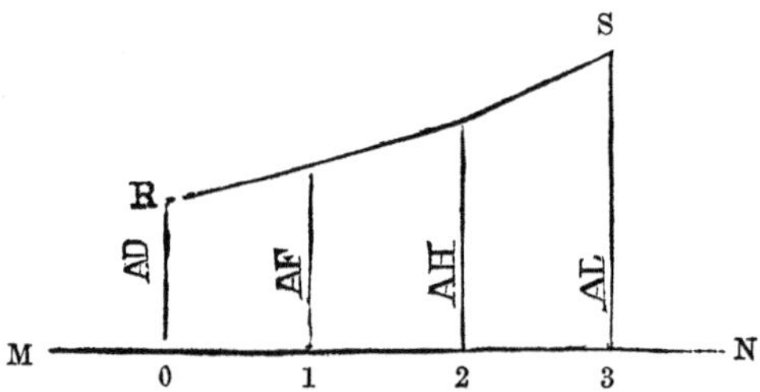

Then shall 0 be the logarithm of AD or 1; 0 1 the logarithm of AF; 0 2 of AH; 0 3 of AL; and so on.

If the distance between the points 0, 1, 2, 3, &c. were indefinitely small, then would the line RS, connecting the extremities of the perpendiculars, become a curve, called the logarithmic curve, and from this might every number of the slide-rule be readily obtained. For practical purposes, however, we shall not require distances less than the eighth of an inch: but it will be advisable to determine the lengths of the perpendiculars by a more tedious process than the one described; as in this, the least error in drawing the parallels is so increased by the divergence of the lines forming the angle, as most frequently to render the curve altogether useless.

From the nature of the lines AD, AF, AH, it follows that AF is a mean proportional between AD and AH. Hence, if the lengths AD and AH were given, AF might be inserted, by problem 14, on the parallel ruler. Thus, to raise a mean proportional between a b and e f,—

Produce e f, and make f g = a b; bisect e g in h, and from the point h, with the distance h g, cut off f k. Bisect b f in m, and make m n = f k. m n is the mean proportional required.

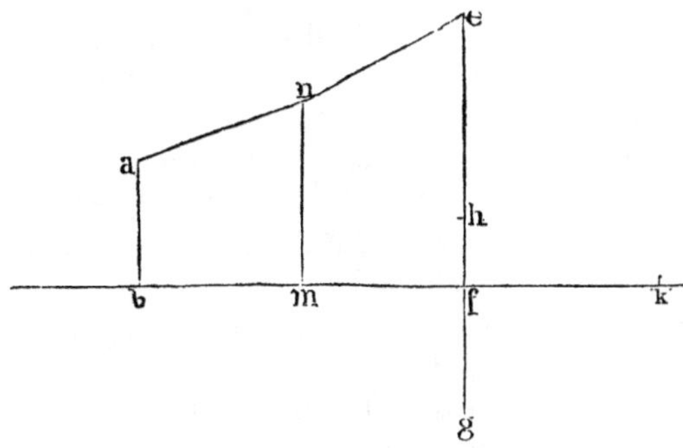

In the same way might a mean proportional be placed

halfway between m and f, and again between f and this, and so on to any degree of exactness.

Instead of bisecting the line b f continually, it will be better to set up a number of perpendiculars first; and, for this purpose, it will be necessary to choose some number represented by $2^n + 1$, as $2^4 + 1$, $2^5 + 1 = 17, 33$, &c., as by this means there will always be a line ready, from which to cut off the mean proportional. Further, in order to obtain ten numbers, the last perpendicular must be taken ten times the smaller. Having, then, set off any distance 16 times along AB, (see title-page, Fig. II.,) and raised 17 perpendiculars, take any height, AC, and complete the parallelogram, ACDB. Divide DB into ten equal parts in the points I, II, III, IV, &c. Between B 1 and AC, find the mean proportional f; between this and AC, g; and so on, till all are determined. Connect their extremities by the logarithmic curve C m a 1, and through the points IX, VIII, VII, VI, &c. draw lines parallel to CD, to meet it in m, k, h, e, &c.; from which points draw the lines m 9, k 8, &c. parallel to AC. The distances from D, toward C, will represent the logarithms of the numbers 2, 3, 4, &c. Lay this line over again to E, and make EF equal to ED; join FC, and draw the parallels; so shall EF be divided logarithmically; and, since EF is twice CD, the numbers on CD shall be the squares of those on EF; and if CG be made equal to one-third of EF, then shall the numbers on CG be the cubes of those on EF, and the square roots of the cubes of those on CD.—Thus, to find the square and cube of 5. Take, with the compasses, the distance from F to 5; this will reach from D to 25, the square, and from G to 125, the cube.—To find the square root of 16. Take 16 from

D, it will reach from F to 4.—To find the square root of 4 cubed. Measure from D to 4; the distance will reach from G to 8.

The subdivision of the portions into tenths is easy; thus, for instance, on the line E F: measure the distance from 1 to 2; this will reach back from 3 to 1½; also from 5 to 2½, from 7 to 3½, and so on: also the distance from 1 to 3, on DC, will reach forward from 4 to 12, from 5 to 15, from 6 to 18, &c.; the distance from 1 to 4 will reach forward from 4 to 16, from 6 to 24, from 7 to 28, &c.; the distance from F to 2 will reach forward from G to 8, from 8 to 64, &c.; and thus, by a little contrivance, may the whole of the subdivisions be filled up.

By means of the logarithmic curve we may double a cube or globe. Thus, suppose the diameter of a globe, or the side of a cube of gold, is half an inch; it is required to find the diameter of another that shall contain twice as much.

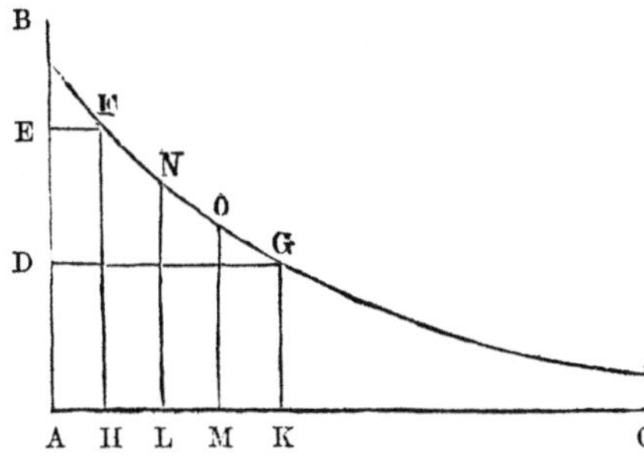

Draw AB at right angles to AC, and make AD equal to half an inch, and AE the double of it. Draw EF and DG parallel to AC, meeting the curve in F and G, from which points let fall the perpendiculars FH, GK. Divide HK into 3 equal parts, and make NL and OM parallel to FH. OM is the diameter of the globe required.

For, from the nature of the curve,

$$KG : OM :: OM : NL :: NL : FH;$$
$$\therefore KG^3 : OM^3 :: KG : FH; \text{ that is, as } 1 : 2.$$

The duplication of the cube is a problem famous in antiquity. It was proposed, by the oracle at Delphi, as a means of stopping the plague which was then raging at Athens.

To lay the numbers down on the rule, however, correctly, we must have recourse to a table of logarithms, as was shown in describing the Sector. The line intended to be numbered is to be divided into 1000 equal parts; then the distance from 1 to 2 will be 301 of those parts, this number being the logarithm of 2; the distance from 1 to 3 will be 477, the logarithm of 3, &c.

THE SLIDE-RULE.

THE Slide-rule is an instrument containing the logarithmic lines we have been describing; they are arranged in different ways, according to the purpose for which they are intended; but the most extensively useful is that in which the D line commences with unity. The line A is laid down twice along the top of the rule; the line D once in the same space, at the bottom of the rule; between them is a groove for the reception of the slide BC, which is merely a copy of the A line. In the rules I have constructed there are two other grooves, for containing two extra slides, when not in use. One of these is marked E, and contains the logarithmic line repeated thrice. The third is a trigonometrical slide, and is graduated with logarithmic sines and tangents, the former of which work to the line D, and the latter to A. At the back is a comprehensive table of numbers, suited to the variety of lines, surfaces, and solids usually met with. In making use of the rule, it is to be observed, that the values of the numbers in the second division of A, B, and C, are ten times those of the first. If the first series be reckoned as .01, .02, .03, &c., the second will be .1, .2, .3, &c. The first .1, .2, .3, &c., the second 1, 2, 3, &c. The first 1, 2, 3, &c., the second 10, 20, 30, &c. And here it is immaterial whether a number is chosen from the first or second division; but, in ascertaining the squares and square roots of numbers with C and D, it will be necessary to observe, that if the number of figures representing the square be odd, or a decimal having an odd number of ciphers before the first effective figure, it must be selected from the first division of C; if otherwise, from the second division. This is analogous to the caution requisite in extracting the roots of numbers by computation, where it is necessary to make the first point fall upon the unit, and to have an even number of figures in the decimals. A little practice on the rule will soon render this familiar: thus, to find the square soot of 5, or .05, look under the 5 in the first division of C; for the square root of 50, .5, or .005, under 5 in the second division of C; for in common arithmetic it is necessary to put .5 into the shape of .50, and .005 to .0050, before we take the root; and the same form, of course, applies to the slide-rule. A like principle applies to the E slide. Whole numbers containing one figure are in the first division, two in the second, and three in the third; and decimals are managed accordingly. Besides this, there is a mutual relation between the lines, which will be readily understood by attending to the remarks that follow.

When the slide BC is laid *evenly* in the groove, that is, when the commencement of A coincides with the commencement of B, the numbers on A are the *same* as the numbers on B; when the slide is in *any other* position, the numbers on A are *proportional to* the numbers on B. The same is the case with the D line. When the slide lies evenly in, the numbers on C are the squares of those on D; when the slide is in any other position, the numbers on C are *proportional to* the squares of those on D. Thus, let the slide be drawn back till the 2 of B falls under the

1 of A, then we have a series of continued proportionals, or equivalent fractions, the odd terms or numerators standing above the even terms or denominators, as $\frac{1}{2}=\frac{2}{4}=\frac{3}{6}=\frac{4}{8}=\frac{5}{10}=$&c., or 1 : 2 : : 2 : 4 : : 3 : 6 : : 4 : 8 : : 5 : 10 : : &c.; and on the C and D lines $\frac{2}{1^2}=\frac{8}{2^2}=\frac{18}{3^2}=\frac{32}{4^2}=$ &c., or 2 : 1^2 : : 8 : 2^2 : : 18 : 3^2 : : 32 : 4^2 : : &c. An attention to the above will explain every operation. Whenever we require to square a number, we select this number on D, and look *over* it on C; whenever we wish to obtain the square root of a number, we select the number on C, and look *under* it on D. When we neither require to square it, nor to extract the root, the D line does not enter into the operation, and the A and B lines are used indifferently. We may now proceed to the mode of valuing the numbers.

Let it be required to multiply 3 by 5. Under the 3 of A set the beginning 1 of the BC slide; then over the 5 is 15. Here the number chosen from the first division consisting of a unit, the beginning of B also a unit, and the result falling in the second division, it will consist of tens.

To multiply 3 by 50. Let the slide stand as before; then, the 5 being taken as 50, the beginning of the slide will be 10; therefore the 3 standing over it, in the first division of A, becomes 30; consequently the result, falling in the second division, will be 150.

To multiply 30 by 50. Let the slide remain; the product is now 1500; for the 5 on B being reckoned 50, the commencement of the slide will be 10; therefore the 30 standing over it in the first division becomes 300: consequently the result falling in the second division, the primary number will be thousands.

In dividing numbers, the following considerations are to be attended to. If the divisor contain as many figures as the dividend, the beginning of the line containing the dividend will be unity. If the divisor have *one more* figure, the beginning of the line from which the dividend is chosen will be in the *first* place of decimals; if it have *two* more, in the *second* place of decimals; if *three* more, in the *third* place of decimals; and so on. Thus, divide 4 by 8, that is $\frac{4}{8}=?$ Under 4 of A in the second division place 8 of B, denominator under numerator as in vulgar fractions, then over the beginning of B is .5. For the divisor containing only as many figures as the dividend, the beginning of the second division of A will be 1, and the quotient falling in the first division, it will be .5.

Divide 4 by 80, that is $\frac{4}{80}=?$ Let the slide remain; here, the divisor containing *one* more figure than the dividend, the beginning of the second division will be in the *first* place of decimals, that is .1, consequently the quotient will be .05.

Divide 4 by 800, that is $\frac{4}{800}=?$ Let the slide stand; here, the divisor containing *two* more figures than the dividend, the beginning of the second division will be in the *second* place of decimals, that is .01, consequently the quotient will be .005.

The numbers on C being the squares of those on D, if the beginning of D is 10, that of C will be 100.

Divide by 40, 3 times 8 squared, that is $\frac{3 \times 8^2}{40} = ?$ To 40 on A set 3 of B, over 8 of D is 4.8. Here, 40 containing one more figure than the 3, the 3 becomes .3, consequently the result, falling in the next division, will be 4.8.

Divide by 40, 3 times 80 squared, that is $\frac{3 \times 80^2}{40} = ?$ Let the slide remain; here the 3 becomes .3 as before, but the commencement of D being reckoned as 10, the numbers on C are increased 100 times, and therefore the .3 becomes 100 times .3 or 30; hence the result, falling in the next division, will be 480.

To divide by decimals, add as many ciphers to the dividend as the first effective figure is removed from the decimal point; thus $\frac{10}{.2} = ?$ To 10 of A place .2 of B; the 2 is in the *first* place of decimals, therefore add *one* cipher to the 10 and it becomes 100. Look back to the beginning of the slide, and over it is 50.

What is the value of $\frac{16}{.04}$? To 16 of A set 4 of B; the 4 being the *second* place of decimals, add *two* ciphers, and we have 1600 for the value of the 16. Look back to the beginning of the slide, over which is 400.

Find the value of $\frac{3 \times 40^2}{.08}$. To 8 on A set 3 of B, over 40 of D is 60,000. Here the 8 of .08 being the *second* place of decimals, add *two* ciphers, and the 3 becomes 300,

(3 with two ciphers.) Again, the beginning of D being 10, the numbers on C become increased by 10^2, or 100, so that two more ciphers have to be added, and the 3 becomes 3 with four ciphers, that is 30,000; consequently the result falling more to the right will be 60,000.

Find the value of $\frac{7 \times 40^2}{60^2}$, that is, divide by 60^2, 7 times 40^2. To 60 of D set 7 of C, over 40 is 3.111. Here 60 and 40, the two numbers selected on D, having each the same number of figures, the result falling on C will contain the same number of integers as the 7; hence the quotient is 3.111.

Divide by 60^2, 7 times 4^2, that is $\frac{7 \times 4^2}{60^2} = ?$ Let the slide stand as before. Here the 4 of the D line containing *one* figure less than the 60, the square of it will contain *two* figures less, consequently the 7 over the 60 is reduced to the second place of decimals, and becomes .07; hence the result will be .03111.

Divide by 20^2, 70 times 50. Over 20 on D set 70 of the slide, and look under 50 of A. Here, the commencement of the D line being 10, the beginning of the A line would be 100; but if we select 50 from the first division of A, we reduce it one-tenth, and the 70 becomes 7; consequently the number under 50 of A is 8.75.

These exercises, if carefully attended to, will be amply sufficient to enable the student to value all quantities correctly.

A very common operation on the slide-rule is to find a mean proportional between two numbers, that is, to extract the square root of their product. Let the two numbers be 4 and 9. Multiply 4 by 9, and take the root; that is, to 4 of A set commencement of slide, under 9 is 6. As the slide stands, it necessarily follows that 4 of the slide is over 4 of D, since 4 is the square root of 4 times 4; hence, in finding the mean proportional between two numbers, it may be effected with the C and D lines only, by setting one of the numbers over itself, and looking under the other. Thus, what is the mean proportional between 3 and 12? Set 3 over 3, then under 12 is 6. The object intended to be accomplished by finding this mean proportional, is to reduce parallelograms to squares, and ellipses to circles; a square whose side is 6 inches being equal to a parallelogram 12 inches by 3, and an ellipse, whose axes are 12 and 3, equal to a circle whose diameter is 6.

As we shall have occasion hereafter to introduce various formulæ for solids, it will be necessary for the learner to study the following operations:—

Find the value of $\frac{8\,(3^2 + 5^2)}{6^2}$. Now, this is equal to $\frac{8 \times 3^2}{6^2} + \frac{8 \times 5^2}{6^2}$. Therefore it must be effected by obtaining the quotients separately, and then adding them together. Hence, over the 6 of D set 8 of the slide, then—

Over 3 is 2.
Over 5 is 5.55

7.55

Find the value of $\frac{40\,(24^2 + \overline{2 \times 32^2})}{33^2}$. Over 33 of D set 40, then—

Over 24 is 21.2
Over 32 is 37.6
Ditto 37.6

96.4

Find the value of $\frac{40\,(24^2 + 32^2 + 59.2^2)}{46^2}$. Over 46 of D set 40, then—

Over 24 is 10.9
32 is 19.35
59.2 is 66.2

96.45

It sometimes happens that we require to multiply three numbers together. This cannot be done by the kind of rule we have been considering in one operation, but it may be effected by dividing by the reciprocal of one of the numbers. Thus, let it be required to find the product of $4 \times 7 \times 8$. The reciprocal of 4 is $\frac{1}{4} = 25$; hence we have to divide by .25, 7 times 8. To .25 on A set 7 of the slide, under 8 on A is 224. By inverting the slide, and pushing it *evenly* in, that is, until the *end* is under the *beginning* of A, it will be seen that the numbers on B are the reciprocals of the corresponding ones on A; hence if instead of the D line, a line similar to A were laid down under the slide, in an inverted position, it would furnish a series of reciprocals, and then three numbers might at once be multiplied together, by taking one of them on this inverted line, one on the slide, and the third on the A line, under which would be the product. Moreover, if, instead of laying it down so as to make the *commencement* of it fall under the end of the slide, it were drawn out toward the right hand till *some other number than unity* stood under the end of the A line, then the product of the three numbers would be divided by this constant number. For instance, suppose we wished to divide by 2, 7 times 8 times 6; that is, to find the value of $\frac{7 \times 8 \times 6}{2}$ Let the inverted line be placed so that the 2 shall fall under the end of the A line; then over the 7 of this inverted line place 8 of the slide, and under the 6 of A will be 168. Hence, if a person pursued an occupation in which his calculations required to be divided by a *constant number*, he might have a rule constructed to suit himself for that particular number. A few such rules are in use. The officers of the customs have frequently to measure pieces of timber, the length of which is taken in feet, and the breadth and thickness in inches. Now, multiplying these three dimensions together, and dividing by 144, gives the solidity in cubic feet. Hence, let the A, B, and C lines be laid down as usual, and instead of D substitute an inverted A line, so placed that 144 shall fall under the end of the slide. Then, if a piece of timber measures 55 feet long, 24 inches broad, and 9 thick; under 55 of A place 24 of the slide, and over 9 of the inverted line is 82½ cubic feet, the content. In malt gauging again, the number of cubic inches in a bushel is 2218.19. Hence, taking the dimensions in inches, let the inverted line be so placed that the number 2218.19 shall fall under the end of the slide;

then if a cistern of malt measures 30 inches long, 16 broad, and 12 deep; to 30 on A set 16 of the slide, and over 12 of the inverted line is 2.6 nearly, the content in bushels. If we wish to obtain the result in gallons, (as 8 gallons make a bushel,) take 8 times one of the dimensions: for instance, to 240 on A set 16 of the slide, and over 12 of the inverted line is 20.78 gallons.

In practice these rules are of the utmost convenience possible, and the principle might be carried out with advantage to a much greater extent than it yet has been.

OBSERVATIONS.

There are three kinds of measure—lineal, superficial, and solid: lineal, for such things as have length only; superficial, for those that have length and breadth; and solid, where there are length, breadth, and thickness. When *lines* vary proportionally they vary simply as their measures; when *surfaces* vary proportionally they vary as the *squares* of their like measures; and when *solids* vary proportionally they vary as the *cubes* of their like measures. Thus, let there be two similar funnels, or cones, A and B; and let A be filled with water to the depth of 1 foot, and B to the depth of 2 feet; then the circumference of the top of the water in B will be twice that of A, both being *lines;* the *area* of the top of the water in B will be 4 times that of A, or 2^2, both being *surfaces:* and the *weight* or *quantity* of the water in B will be 8 times that of A, or 2^3, both being *solids:* and so of all surfaces and solids that vary proportionally.

If a number of regular polygons have equal perimeters,

that contains the greatest amount of surface in which the perimeter is distributed among the greatest number of sides; and, as a circle may be conceived to be a polygon of an *infinite* number of sides, *it* therefore contains the greatest quantity of space within the shortest bounding line.

A regular polygon contains more than an irregular polygon of the same number of sides, their perimeters being equal; thus, an equilateral triangle has a greater area than any other triangle of equal ambit; and a square is the largest quadrilateral that can be constructed with sections of the same line.

In the same way as the circle contains the largest surface within the least compass, so the sphere contains the greatest bulk within the smallest space.

RATIOS AND GAUGE POINTS.

At the back of the rule will be found a quantity of tabular work, adapted to various kinds of calculation: these consist of ratios and gauge points. Ratios express the proportions existing between certain lines, or numbers; thus, if the diameter of a circle be 113, its circumference will be 355; and, as the circumference varies as the diameter, therefore 113 : 355 expresses the ratio of the diameter of any circle to its circumference. Gauge points are the square roots of divisors; thus, if we require to reduce square inches to square feet, we must divide by 144, which number may be chosen on A; if instead of this we divide by 12^2, we take 12 upon the D line, and, for the sake of distinction, 12 is called a gauge point. In rules having the D line commencing with unity, when the slide is set to any gauge point, it is also set to the corresponding divisor, the one standing under the slide, the other above it; and therefore, with such rules, it would be immaterial whether we used divisors or gauge points; as however, the formulæ for many surfaces, and almost all solids, require the use of the D line, it is far more convenient for valuing the numbers, to make use of gauge-points, and therefore the tabular work is so constructed.

TABLE I. contains a list of ratios belonging to the circle, commencing—

A		B
113 diameter	=	355 circumference.
44 diameter	=	39 side of equal square.

That is, under 113 on A set 355 on B, then the numbers on A will be a series of diameters, and the numbers beneath them on B will be their corresponding circumferences; and so of all the rest.

EXAMPLES.

1. If the diameter of a circle is 8 inches, what is its circumference? Set the rule as directed, then under 8 is 25.13 inches.

2. The diameter of a circle is 9 inches, what is the side of an equal square? Under 44 of A set 39 of B; under 9 is 7.97 inches.

3. The radius of a circle is 6 inches, what is the length of an arc of it containing 31½ degrees? Under 57.3 degrees on A set 6 inches on B; under 31½ degrees on A is 3.3 inches.

4. The circumference of a circle is 75, what is the diameter?—Ans. 23.87.

5. The diameter is 7, what is the circumference?—Ans. 22 nearly.

6. The diameter is 17, what is the circumference?—Ans. 53.4.

7. Suppose the diameter of the earth to be 7960 miles, what is its circumference?—Ans. 25,000 miles.

8. The diameter of a circle is 6 inches, what is the side of a square inscribed within it?—Ans. 4.24.

9. The circumference of a circle is 12 feet, what is the side of an equal square?—Ans. 3.38.

10. The circumference of a circle is 15 inches, what is the side of its inscribed square?—Ans. 3.375.

11. The side of a square is 10 inches, what is the diameter of an equal circle?—Ans. 11.28.

12. The side of a square is 20 yards, what is the circumference of an equal circle?—Ans. 70.83.

13. The side of a square is 19 inches, what is the side of an equal equilateral triangle?—Ans. 28.88.

14. The area of a circle is 27, what is the area of a square inscribed in it?—Ans. 17.18.

15. An arc of 38 degrees measures 5 inches, what is the radius of the circle of which it is a part?—Ans. 7.56.

16. I have a circular piece of wood, whose diameter is 15 inches, and wish to cut the largest square out of it; what will be the length of each side?—Ans. 10.6 inches.

The method of obtaining these ratios in whole numbers is a beautiful exemplification of the abridgment of labour effected by the slide-rule; and of performing, with the utmost facility, operations that would require considerable time and trouble by any other means. Archimedes discovered that the ratio of 7 to 22 nearly expressed that of the diameter of a circle to its circumference. Purbachius, in the fifteenth century, making the diameter 120, reckoned the circumference at 377. Metius, two centuries later, subtracted the 7 and 22 from the 120 and 377, and obtained the numbers 113 and 355. This last ratio is easily remembered, from its containing the first three odd numbers in pairs, and it is remarkably accurate, the quotient of 355 by 113 being true to the sixth place of decimals. The obtaining of these ratios *in integers*, however, must have been a task of considerable labour. To determine them by the slide-rule is the work of a moment. By various modes of computation it may be shown that if the diameter be 1, the circumference will be nearly 3.1416; therefore, under 1 of A set 3.1416, as nearly as possible, and run the eye along until you find two numbers coinciding: such will be 113 and 355, which will be the ratio required. The advantage of having the ratios in whole numbers, for the purposes of the slide-rule, is obvious, as they can be set with greater rapidity and exactness than decimals.

The following table will enable the student to solve the previous questions numerically:—

Diameter 1, circumference 3.1416.
———— side of equal square, .8862.
———— side of inscribed square .7071.
Circumference 1, diameter .3183.
———— side of equal square .2821.
———— side of inscribed square .2251.
Side of square 1, diameter of equal circle 1.128.
———— circumference of equal circle 3.545.
———— side of equal equilateral triangle 1.5196.
Area of circle 1, area of inscribed square .6366.
Area of square 1, area of inscribed circle .7854.
———— area of inscribed octagon .8284.
The length of an arc of 57.2957795 degrees = radius of circle.

Solution of question 8:—.7071 × 6 = 4.2426.

TABLE II. contains the Linear Dimensions of Polygons described within and without Circles, and commences thus:—

No. of Sides.	Inscribed Polygon.		Circumscribed Polygon.	
	A. Diam.	B. Side.	A. Diam.	B. Side.
3	15	13	15	26
4	9.9	7	1	1

That is, if the diameter of a circle be 15, the side of an equilateral triangle inscribed within it will be 13: hence, under 15 of A set 13 of B; then the numbers on A will be a series of diameters, and the numbers beneath them on B will be the sides of the corresponding triangles; and so of the rest. The method of obtaining them is first by computation, and then as for the ratios before described.

EXAMPLES.

17. The diameter of a circle is 12 inches, what will be the side of an equilateral triangle inscribed therein?—Under 15 of A set 13 of B; under 12 of A is 10.4 inches.

18. The diameter of a circle is 11½ inches, what is the side of a regular pentagon inscribed within it?—Ans. 6.76.

19. A circle whose diameter is 9¼ inches has a regular hexagon surrounding it, what is the length of each side?—Ans. 5.33.

20. A person having a circular piece of ground 37 yards in diameter, wishes to make within it a flower-bed of a heptagonal form, whose area shall be a maximum; what will be the length of each side?—Ans. 16.

21. If I make the diameter of a circle a parallel distance on the line L of the sector from 100 to 100, what parallel distance must I take off as the side of an undecagon inscribable therein?—Ans. 28.1.

The following table will enable the student to solve the preceding questions numerically.

The diameter of the circle being unity,

No. of Sides.	Side of Inscribed Polygon.	Side of Circumscribed Polygon.
3	.8660254	1.7320508
4	.7071068	1.0000000
5	.5877853	.7265425
6	.5000000	.5773503
7	.4338837	.4815745
8	.3826834	.4142136
9	.3420201	.3639702
10	.3090170	.3249197
11	.2817325	.2936264
12	.2588190	.2679492

Solution of question 20:—.4338837 × 37 = 16.0536969.

TABLE III. contains the Areas of Polygons, commencing thus:—

No. of Sides.	C. Area.	D. Side.
3	3.9	3
5	43	5

That is, if 3 be the side of an equilateral triangle, its area will be 3.9, and, as similar surfaces vary as the squares of their like measures, if over 3 of D we set 3.9 on C, then the numbers on D will be a series of sides, and the numbers over them on C their corresponding areas.

EXAMPLES.

22. The side of an equilateral triangle is 2, what is its area? Over 3 of D set 3.9 on C; over 2 of D is 1.732, the area required.

23. Required the area of a regular nonagon having a side of 7.3 yards.—Ans. 329 yds.

24. What is the area of an undecagon whose side measures 6.4 feet?—Ans. 383.6 feet.

25. The side of an octagon is 4.9 feet, what is its area? —Ans. 116 nearly.

The side being given in inches, to find the area in square feet, take 12 times the number on D, for the number of inches in a square foot is 12^2.

EXAMPLES.

26. The side of an equilateral triangle is 19 inches, how many square feet does it contain?—Over 36 of D set 3.9 of C; over 19 of D is 1.086 feet.

27. The side of a regular pentagon is 53 inches, how many square feet does it contain?—Ans. 33.55.

28. What is the area of a nonagon, each of whose sides measures 27 inches?—Ans. 31.29.

The side being given in feet, to find the area in square yards, take 3 times the number on D, for the number of feet in a square yard is 3^2.

EXAMPLES.

29. The side of a regular pentagon is 7 feet, how many square yards does it contain?—Over 15 of D set 43 of C; over 7 of D is 9.36 yards.

30. What is the area of a heptagon whose side measures 17 feet?—Ans. 116.7 yards.

31. A decagon measures 20.2 feet along each side, what is the area?—Ans. 348.8 yds.

The following table will enable the student to solve the preceding questions numerically. Multiply the subjoined numbers by the square of the side.

Equilateral Triangle	.4330127
Pentagon	1.7204774
Hexagon	2.5980762
Heptagon	3.6339124
Octagon	4.8284272
Nonagon	6.1818242
Decagon	7.6942088
Undecagon	9.3656411
Dodecagon	11.1961524

Solution of Question 22:—$.4330127 \times 2^2 = 1.7320508$.

FALLING BODIES.

TABLE IV. is of a miscellaneous nature, and will be understood by inspection. It is found that a heavy body, in the latitude of London, falls 579 feet, or 193 yards in 6 seconds; and the spaces descended by falling are as the squares of the times; hence, as directed by the table, over 6 seconds on D set 579 feet on C, (or 193 yards, if the distance be required in yards,) then the numbers on C will be a series of distances fallen, and the numbers beneath them on D the seconds elapsed in falling. The same law applies to bodies projected directly upwards, the retardation corresponding with the acceleration in an inverse order.

EXAMPLES.

32. How many feet will a body fall in 1 second?—Over 6 of D set 579 on C; over 1 of D is $16\frac{1}{12}$ feet.

33. If a ball is propelled straight upwards, and is found to be 18 seconds before it again falls to the earth, how many yards has it ranged?—9 seconds occupied in ascending, 9 in descending; over 6 of D set 193 on C; over 9 is 434 yards.

34. Standing at the mouth of a well, which, by means of reflecting the sun's rays into it with a mirror, I perceived to be of considerable depth, I dropped a stone into it, and found it reached the water in $3\frac{1}{2}$ seconds; what was its depth?—Ans. 197 feet.

35. How long would a cannon-ball, fired perpendicularly upwards, be in rising a mile, if it went no higher?—Ans. 18.12 seconds.

PENDULUMS.

A pendulum 22 inches long, as shown by the table, makes 80 vibrations, or 40 revolutions per minute, and their lengths vary reciprocally as the squares of their times, their velocity being regulated by the force of gravity, like that of falling bodies. Hence, invert the slide, and set 22 inches on B over 80 vibrations, or 40 revolutions on D; then the numbers on the inverted line will be a series of lengths, and the numbers beneath them on D the corresponding number of vibrations or revolutions.

EXAMPLES.

36. What is the length of a pendulum vibrating 60 times per minute?—Over 60 is 39.2 inches.

37. What is the length of a pendulum vibrating 64 times per minute?—Ans. 34.4 inches.

38. What is the length of a pendulum making 29 revolutions per minute?—Ans. 42 inches.

EXPERIMENT.

Suspend from a hook in the ceiling a string with a bullet at the end; set it vibrating, or swing it round so as to cause it to revolve, and compare its motions with a watch.

The next part of the tabular work relates to the areas of circles and surfaces of spheres, and is as follows:—

Circle 43 area C = 7.4 diameter D.
23 area C = 17 circumference D.
Sphere 172 surface C = 7.4 diameter D.
92 surface C = 17 circumference.

The student will perceive from this that the surface of a sphere is 4 times the area of a circle of equal diameter. That is, if an orange were perfectly round, and cut into two equal parts, then the external surface of the rind in each half would be just double the surface of the part cut by the knife. Similar surfaces varying as the squares of their like measures, the dimensions being taken on D the areas will be on C.

EXAMPLES.

39. The diameter of a circle is 5 inches, what is its area?—Over 7.4 of D set 43 of C; over 5 is 19.63 square inches.

40. What is the area of a circle whose circumference is 12 inches?—Ans. 11.46 inches.

41. The circumference of a sphere is 12 inches, what is the surface?—Ans. 45.8 square inches.

If the dimensions are given in inches and the area is required in feet, take 12 times the number on D; and if the dimensions are in feet, and the area is required in square yards, take 3 times the number on D.

EXAMPLES.

42. The diameter of a circle is 19 inches, what is its area in square feet? 12 times 7.4 = 88.8; over 88.8 on D set 43 on C; over 19 is 1.97 square feet.

43. The circumference of a circle is 43 inches, how many square feet does it contain?—Ans. 1.021 square feet.

44. The diameter of a sphere is 17 feet; what is its surface in square yards?—Ans. 100.8 square yards.

The next part of the table is—

1 C = side of square or cube D.
2 C = diagonal of square or diameter of circumscribing circle D.
3 C = diagonal of cube or diameter of circumscribing sphere D.

That is, set 1 of C over the side of a given square on D, then under 2 of C will be its diagonal, or the diameter of its circumscribing circle.

EXAMPLES.

45. A circle 12 inches in diameter has a square inscribed within it, what is the length of each side?—Over 12 of D set 2 of C; under 1 of C is 8.48 inches.

46. A cube measures 7 inches along the side, what will be the diagonal of the face, and what of the cube?—

Ans. 9.9 diagonal of the face.
12.12 diagonal of the cube.

47. What is the longest line that can be taken in a cubical box whose sides measure 19.4 feet?—Ans. 33.6 feet.

48. A square inscribed in a circle measures 43 inches, what is the diameter of the circle?—Ans. 60.8 inches.

49. The diameter of a sphere is 26.7 inches; what will be the side of the largest cube that can be cut from it?—Ans. 15.41.

50. Standing within a cubical room I found that the distance from one of the top corners to the opposite corner at bottom was 23.3 feet; what was the distance of the ceiling from the floor?—Ans. 13.45 feet.

VELOCITY OF SOUND.

THE flight of sound is uniformly proportional to the time; hence use the A and B lines as directed in the table.

EXAMPLES.

51. I observed the flash of a gun 12 seconds before hearing the report; how far was it distant from me?—To 65 seconds on A set 14 miles on B; under 12 of A is 2.59 miles on B.

52. I observed a flash of lightning, and 7 seconds afterwards heard the thunder; how far distant was the electric cloud?—Ans. $1\frac{1}{2}$ mile.

53. A person standing on the bank of a river, heard the echo of his voice reflected from a rock on the opposite bank in 5 seconds after; what was the breadth of the river?—Ans. 950 yards.

The subjoined table will enable the student to solve the questions by computation:—

A body falls $16\frac{1}{12}$ feet in the first second.

A pendulum vibrating seconds in the latitude of London, is 39.1396 inches.

In a pendulum describing a conical surface, the time of revolution is equal to the time of two oscillations of a simple pendulum, equal to the height of the cone; that is, a pendulum takes the same time in going half round a circle as it does in falling across it.

Putting *d* diameter, *c* circumference,

Area of Circle = $.7854\ d^2$ or $.07958\ c^2$.
Surface of Sphere = $3.1416\ d^2$ or $.31832\ c^2$.

Putting s side of square or cube, then diagonal of square, or diameter of circumscribing circle $= s\sqrt{2} = s \times 1.4142136$.

Diagonal of cube, or diameter of circumscribing sphere $= s\sqrt{3} = s \times 1.7320508$.

Sound flies about 380 yards per second.

Solution of Question 33 .$16\frac{1}{12} \times 9^2 = 1302\frac{3}{4}$ feet $= 434\frac{1}{4}$ yards.

SURFACES.

We now come to Table 5, which consists of a number of gauge points for the mensuration of surfaces, quadrilaterals, triangles, parabolas, circles, cycloids, and ellipses; and the surfaces of prisms, cylinders, pyramids, cones, and spheres. The area of a rectangle is equal to the product of the length and breadth. The area of a trapezoid is found by multiplying half the sum of the parallel sides by the perpendicular distance between them. A triangle is half a rectangle, and therefore its area is half the product of the height and base.* The areas of trapeziums and multilaterals are found by dividing them into triangles. A parabola is equal to $\frac{2}{3}$ of its circumscribing parallelo-

---

* If a quadrilateral can be inscribed in a circle, its area will be, (putting s semiperimeter, and a, b, c, d, the sides,) $= \sqrt{\overline{s-a}.\,\overline{s-b}.\,\overline{s-c}.\,\overline{s-d}}$. If one of the sides, as d, is supposed to vanish, the figure merges into a triangle, and the formula becomes $\sqrt{\overline{s-a}.\,\overline{s-b}.\,\overline{s-c}.\,s}$. That is, for the quadrilateral, from half the sum of the four sides subtract each side separately; multiply the four remainders together; the square root will be the area. For the triangle, from half the sum of the three sides subtract each side separately; multiply the three remainders and the half sum together; the square root will be the area.

gram, and therefore its area is found by taking $\frac{2}{3}$ of the product of the height and base. A circle may be conceived to be a polygon of an infinite number of sides. Now, a polygon is made up of as many triangles as the figure has sides, and the area of each triangle is found by taking the product of the height and half the base; therefore the area of the whole polygon will be equal to the perpendicular multiplied by half the perimeter; this, when the figure merges into a circle, becomes the radius multiplied by half the circumference; or, which is equivalent, the diameter multiplied by $\frac{1}{4}$ of the circumference. Now, when the diameter is 1, the circumference is 3.1416, hence $1 \times .7854 =$ the area of a circle whose diameter is unity. And since similar surfaces are to each other as the squares of their like dimensions, the area of any circle will be equal to the square of its diameter multiplied by .7854. The area of the sector of a circle, in like manner, will be found by multiplying the radius by $\frac{1}{2}$ the length of the arc.* The area of a cycloid is 3 times that of its generating circle. From the method described in page 12, of projecting the circle into an ellipse, it is obvious that the area will be in proportion to the elongation, that is, equal to the product of the axes multiplied by .7854. The sides of a prism being parallelograms, it follows that the perimetrical surface will be equal to the product of

* To find the length of the arc, from 8 times the chord of $\frac{1}{2}$ the arc subtract the chord of the whole arc, and divide by 3; the quotient is the length, nearly.

To find the length of the chord of $\frac{1}{2}$ the arc, add together the square of the versed sine, or height of the segment, and the square of $\frac{1}{2}$ the chord; the square root is the length of the chord of $\frac{1}{2}$ the arc.

the perimeter and height. The same rule applies to the cylinder, which is a round prism. The sides of a pyramid being triangles, the area of each of which is found by multiplying $\frac{1}{2}$ the base by the height, the sloping surface will be found by multiplying $\frac{1}{2}$ the perimeter by the slant height. The same will be the case with the cone, which is a round pyramid. The surface of a sphere is equal to the convex surface of its circumscribing cylinder, which surface, as above shown, is equal to the circumference multiplied by the depth; that is, in this case, the circumference multiplied by the diameter. The same holds good for the surface of any part of the sphere, it still being equal to the surface of the corresponding parallel section of the cylinder.

We may now refer to Table 5, at the back of the rule, the gauge points of which are determined as follows. To reduce square inches to square feet, we divide by 144, and $\sqrt{144} = 12$, the gauge point for squares; $144 \div .7854 = 183.3462$, and $\sqrt{183.3462} = 13.54$ the gauge point for circles, when the dimensions are taken in inches, and the area is required in square feet; and so of the rest. Putting s side, b base, p perimeter, h perpendicular height, H slant height, d diameter, c conjugate axis, t transverse; then

Area of square $= \dfrac{1s^2}{\text{square.}}$ Parallelogram $= \dfrac{bh}{\text{square.}}$

Triangle $\dfrac{\frac{1}{2}bh}{\text{square.}}$ Parabola $\dfrac{\frac{2}{3}bh}{\text{square.}}$

Surface of prism and cylinder $= \dfrac{ph}{\text{square.}}$ Surface of pyramid and cone $= \dfrac{\frac{1}{2}pH}{\text{square.}}$

Area of circle $= \dfrac{1d^2}{\text{circle.}}$ Ellipse $= \dfrac{ct}{\text{circle.}}$

Area of cycloid $= \dfrac{3d^2}{\text{circle.}}$ Surface of sphere $= \dfrac{4d^2}{\text{circle.}}$

Surface of spherical zone $= \dfrac{4dh}{\text{circle.}}$

That is, to obtain the area of a square, over the square gauge point on D, set 1 of the slide, then over the side on D is the area; for the parallelogram over the square gauge point on D, set the base, then under the height on A is the area; or find a mean proportional between the base and height, then over the square gauge point on D set 1, and over the mean proportional on D will be the area; for the parabola, over the square gauge point on D set $\frac{2}{3}$ of the base, then under the height on A is the area; for the surface of a sphere, over the circular gauge point on D set 4 of the slide, then over the diameter on D is the surface; and so of the rest.

EXAMPLES.

54. The diameter of a circle is 17 inches; how many square feet does it contain?—Over 13.54 on D set 1 of the slide, over 17 is 1.576 feet.

55. The base or double ordinate of a parabola is 39 inches, the height or absciss 11.1 inches; what is the area in feet?—$\frac{2}{3}$ of 39 = 26. Over 12 of D set 26 of the slide, under 11.1 of A is 2 feet.

56. The diameters of an ellipse are 20 and 17 feet; what is the area in square yards?—Over 3.385 on D set 20 on the slide, under 17 on A is 29.67 yards; or, to 17

of D set 17 of the slide, under 20 is 18.44, the mean proportional: then, over 3.385 on D set 1, over 18.44 is 29.67, as before.

57. The side of a square measures 17 inches; required the area in square feet.—Ans. 2 square feet.

58. As a wheel, 5 feet in diameter, is rolled along by the side of a wall, a nail, bent sideways over the tire, scratches it, and marks out a succession of curves, termed cycloids; what is the area of each in square yards?—Ans. 6.545 yards.

59. A circular field measures 283 yards in diameter; how many acres does it contain?—Ans. 13 acres nearly.

60. A globe is 7 feet in diameter; what is the extent of its surface in square yards?—Ans. 17.1.

61. A piece of land measures 95 links by 74; how many square perches does it contain?—Ans. 11.24.

62. How many square yards of canvas will be required to construct a conical tent, 57 feet round the bottom, the slant height of which is to be 22 feet?—Ans. 69.7.

63. Required the surface, in square feet, of a pentagonal prism, the length 168 inches, and each side of the base 33 inches.—Ans. 192.5.

64. How many square yards are contained in a parabola, of which the base is 126, and the height 210 inches?—Ans. 13.6.

The following Table of Divisors will enable the student to solve the preceding questions numerically. The same formulæ apply.

Dimensions in	Area in	Square.	Circle.
Inches..........	Sq. Inches......	1	1.2732
	Feet	144	183.3462
	Yards	1296	1650.1164
Feet.............	Yards	9	11.4591
	Rods........	272.25	346.639
Links...........	Perches....	625	795.7737
Yards...........	Perches....	30.25	38.5154
	Acres.......	4840	6162.4719
Chains..........	Acres.......	10	12.7323

Solution of Question 58: $\frac{3 \times 5^2}{11.4591} = 6.545$ square yards.

The following is not adapted for the slide-rule; but as it is an excellent method, and requisite to complete the mensuration of surfaces, it is accordingly inserted.

To find the areas of plain figures by an odd number of equidistant ordinates.

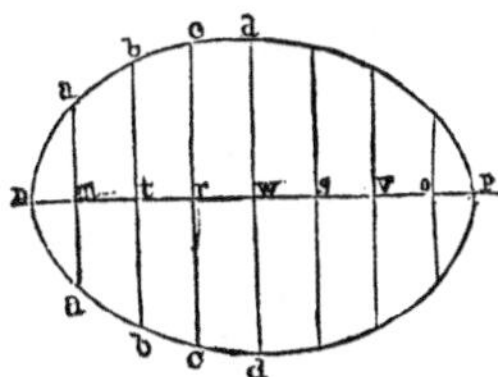

Find the centre of the figure w, and draw the diameters np, dd. On each side of w set off any equal distances ws, sv, vo, wr, rt, tm, as often as may be deemed necessary, and through the points m, t, r, &c., draw the ordinates aa, bb, cc, &c., and measure their lengths; also the distance nm, or op, which are equal to each other.

Place in a line the letters x $4e$ $2o$

(Contractions for the words, extreme, four times even, twice odd.)

Set the first ordinate, aa, under x; the second, bb, under 4 *e*; the third, *cc*, under 2 *o*; the fourth under 4 *e*; the fifth under 2 *o*; the sixth under 4 *e*; and so on, alternately, to the last, which set under *x*. Add up the three columns separately.

Multiply the one under 4 *e* by 4; and the one under 2 *o* by 2.

Add the three together, and multiply by the common distance *ws*.

For the end areas multiply the sum of the extreme ordinates, standing under *x*, by twice the height *nm*; that is, the sum of the bases by the sum of the heights.

Add this product to the other, and divide by 3, gives the area.

EXAMPLE.

65. In a curvilinear figure, 7 ordinates were taken in the following order, 20, 32, 38, 41, 39, 33, 22; the common distance *ws* was 8; the distance *nm* 3: required the area—

x	4 *e*	2 *o*
20	32	38
22	41	39
42	33	77
6	106	2
252	4	154
	424	
	154	
	42	
	620	
	8	
	4960	
	252	
	3) 5212	
	1737⅓ area.	

The greater the number of ordinates taken, the more correct will be the area found; and when the curve *ana* is abrupt, the distance *nm* should be small. If the curve taper gradually the distance *nm* may be taken equal to *mt*, and then the extreme ordinates will be 0. The number of ordinates must always be odd. Beginning with one in the middle insures this.

66. In a curvilinear figure 5 ordinates were taken, 70, 79, 80, 78.6, 69; their common distance was 24; the height of each of the end areas 8: required the area.—Ans. 8176.53.

67. In a triangle the ordinates were 0, 2, 4, 6, 8; the common distance 3; required the area.—Ans. 48.

68. The ordinates in a triangle are 0, 3, 6, 9, 12, 9, 6, 3, 0; the common distance is 4; what is the area?—Ans. 192.

SOLIDS.

THE next part on the Slide Rule is Table 6, which consists of a number of gauge-points for the mensuration of Solids. The content of a prism, or cylinder, is found by multiplying the area of the base by the height. Pyramids and cones are ⅓ of their circumscribing prisms and cylinders, and therefore their content will be equal to the product of the area of the base, and ⅓ of their height. A globe is ⅔ of its circumscribing cylinder, and therefore its content is equal to the area of one of its great circles multiplied by ⅔ of its diameter. The number of cubic inches in a gallon, is 277.274; hence, if the dimensions of a square prism are taken in inches, and the content is required in gallons, it will be (putting l the length, and s the side of the prism) $\frac{l\,s^2}{277.274}$, and $\sqrt{277.274} = 16.65$, the gauge point for square prisms. Since the pyramid is $\frac{1}{3}$ of the prism, we may multiply by the *whole height*, and divide by *three times* 277.274: that is, the content will be $\frac{l\,s^2}{831.822}$ and $\sqrt{831.822} = 28.84$, the gauge-point for square pyramids. A gallon of water weighs exactly 10 lbs.; ∴ 1 lb. occupies 27.7274 cubic inches; dividing this by the specific gravity of any metal, and taking the square root of the quotient, gives the gauge point for such metal. The gauge points for polygonal prisms are obtained by dividing the number of cubic inches in a gallon by the polygonal numbers given at page 118, and taking the square root. In treating of surfaces, it was seen that the area of a circle, inscribed in a square whose side is unity, is .7854. Now, $277.274 \div .7854 = 353.0353$; consequently, the dimensions being taken in inches, as before, the content of a cylinder will be (putting l length and d diameter) $\frac{l\,d^2}{353.0353}$ and $\sqrt{353.0353} = 18.78$, the gauge point for cylinders. In the same way as the pyramid was determined by taking 3 times the prismatic divisor, so the content of the cone will be found by taking 3 times the cylindrical divisor; and $3 \times 353.0353 = 1059.106$; consequently, content $= \frac{l\,d^2}{1059.106}$, and $\sqrt{1059.106} = 32.54$ the conical gauge point. The globe, being $\frac{2}{3}$ of the cylinder, will be twice the cone; hence the divisor will be the half of 1059.106, namely, 529.553;

therefore, putting d the diameter, the content of the globe will be $\frac{d\, d^2}{529.553}$, and $\sqrt{529.553} = 23$, the gauge point for globes. By having divisors and gauge points thus prepared, round solids are reduced to square ones, by which means their contents are determined with the greatest ease, as they all come under the general formula $\frac{l\, s^2}{G}$, in which l represents the length, s the side or diameter, as the case may be, and G the prepared divisor, or, for the purposes of the Slide-rule, its square root, the gauge point.

For finding the solidities of frustums the following is an invaluable rule, and of general applicability :—Find the *area* of the top, the *area* of the bottom, and *four* times the middle area ; their sum is *six* times a mean area, which, being multiplied by *one-sixth* of the depth, gives the content. Now, since by the above-mentioned divisors we have reduced round solids to square ones, the rule becomes : Add together the *square* of the top, the *square* of the bottom, and *four* times the square of the middle, and multiply the sum by *one-sixth* of the depth. But four times the square of a number is equal to the square of twice that number ; therefore the rule becomes still easier. Add together the square of the *top*, the square of the *bottom*, and the square of *twice the middle*, and multiply by one-sixth of the depth. Moreover, when solids do not bulge in the middle, like globes and spindles, but taper regularly like cones and pyramids, then the sum of the top and bottom *will be* twice the middle diameter. Therefore, for all regularly tapering frustums the above given rule becomes still more concise, viz. : Add together the square of the *top,* the square of the *bottom,* and the square of their *sum,* and multiply by *one-sixth* of the depth. In the same way as for cones, we multiplied by *the whole* of the height, and took *three times* the divisor, so for frustums we may take the *whole* of the height, and divide by *six times* the divisor. Now, 6 times 277.274 = 1663.644, whose square root = 40.78, the gauge point for square frustums. Also, 6 times 353.0358 = 2118.2148, whose square root = 46, the gauge point for round frustums. Moreover, as a rule that applies to frustums, applies also to the complete solids themselves, and as this is of such general utility, we shall illustrate it by a few examples.

EXAMPLES.

69. A carpenter has a block of wood 11 inches square at top, 13 inches square at bottom, and 12 inches deep: does it contain an exact cubic foot, or more, or less?

Top $11^2 = 121$
Bottom $13^2 = 169$
Sum $24^2 = 576$

866
$2 = \frac{1}{6}$ of depth.

1732 = 4 cubic inches more than a solid foot.

70. A prismoid, 24 inches deep, measures 12 inches by 10 at top, and 16 by 12 at the bottom; what is the content in cubic inches?

Top $12 \times 10 = 120$
Bottom $16 \times 12 = 192$
Sum $28 \times 22 = 616$

$928 \times 4 = 3712$ cubic inches.

71. A wedge measures 8 inches along the edge; the base is 12 inches long, and 4 thick, and the perpendicular height 18 inches; what is the solidity?

Top $8 \times 0 = 0$
Bottom $12 \times 4 = 48$
Sum $20 \times 4 = 80$

$128 \times 3 = 384$ cubic inches.

72. A cylindroid, or solid bounded at one end by a circle 6 inches diameter, and at the other by an ellipse whose axes are 12 and 10, is 24 inches deep; how many gallons will it contain?

Top $6 \times 6 = 36$
Bottom $12 \times 10 = 120$
Sum $18 \times 16 = 288$

$444 \times 4 = 1776$;

and $\frac{1776}{353.0358} = 5.03$ gallons.

73. What is the solidity of a globe whose diameter is 1? See diagram page 48, and suppose E to be halfway between A and F, and then the diameter being 1, A C will be $\frac{1}{2}$, and AE $= \frac{1}{4}$.

$$\therefore \text{E C} = \sqrt{\frac{1}{4} - \frac{1}{16}} = \sqrt{\frac{3}{16}} = \frac{1}{4}\sqrt{3}$$

$\therefore$ twice EC $= \frac{1}{2}\sqrt{3}$ the middle diameter; then—

Square of top $= 0$
Square of bottom $= 1$
Square of twice middle $= 3$

$$\overline{4} \times \frac{1}{6} + .7854 = .5236,$$

the content as determined by other modes.

To return to the slide. Putting l or h, length, height, or depth, according as the solid is considered lying or standing; d and D, less and greater diameter; m, middle diameter, taken halfway between them; r and R, less and greater radius; s and S, less and greater side; q, square root of product, or mean proportional between two dimensions; f, fixed axis; and v, revolving axis; then the capacities of solids will be denoted by the following formulæ:—

1. Prism $= \frac{ls^2}{\text{prism}}$.

2. Pyramid $= \frac{ls^2}{\text{pyramid}}$.

3. Cylinder $= \frac{ld^2}{\text{cylinder}}$.

4. Cone $= \frac{ld^2}{\text{cone}}$.

5. Sphere $= \frac{dd^2}{\text{globe}}$.

6. Spheroid $= \frac{fv^2}{\text{globe}}$.

7. Rectangular Prism $= \dfrac{lq^2}{\text{square prism}}$.

8. Rectangular Pyramid $= \dfrac{lq^2}{\text{square pyramid}}$.

9. Parabolic Prism $= \dfrac{2lq^2}{\text{square pyramid}}$.

10. Elliptic Cylinder $= \dfrac{lq^2}{\text{cylinder}}$.

11. Elliptic Cone $= \dfrac{lq^2}{\text{cone}}$.

12. Paraboloid, or Parabolic Conoid $= \dfrac{\frac{1}{2}hD^2}{\text{cylinder}}$,

or $\dfrac{2hR^2}{\text{cylinder}}$.

13. Hyperboloid, or Hyperbolic Conoid

$$= \frac{h(D^2 + \overline{2m}|^2)}{\text{round frustum}}, \text{ or } \frac{h(R^2 + m^2)}{\text{globe}}.$$

14. Parabolic Spindle $= \dfrac{lD^2}{\text{globe}} \times .8$

15. Spindles in general $= \dfrac{l\,(D^2 = \overline{2m}|^2)}{\text{round frustum}}$,

or $\dfrac{l\,(R^2 + m^2)}{\text{globe}}$.

16. Frustum of Pyramid $= \dfrac{l\,(s^2 + S^2 + \overline{s + S}|^2)}{\text{pyramidal frustum}}$.

17. Frustum of Cone $= \dfrac{l\,(d^2 + D^2 + \overline{d + D}|^2)}{\text{round frustum}}$.

18. Frustum of Paraboloid $= \frac{\frac{1}{2}h(d^2 + D^2)}{\text{cylinder}}$,

or $\frac{2h(r^2 = R^2)}{\text{cylinder}}$.

19. Frustum of Hyperboloid $= \frac{h(d^2 + D^2 + \overline{2m}|^2)}{\text{round frustum}}$,

or $\frac{h(r^2 + R^2 + m^2)}{\text{globe}}$.

20. Middle Frustum of Parabolic Spindle

$= \frac{l(d^2 + 2 \,.\, D^2 - \frac{1}{10}\,(\text{of 2 diff.})^2)}{\text{cone}}$.

21. Middle Frustum of Spindles in General

$= \frac{l(d^2 + D^2 + \overline{2m}|^2)}{\text{round frustum}}$, or $\frac{l(r^2 + R^2 + m^2)}{\text{globe}}$.

22. Middle Frustum or Spheroid $= \frac{h(d^2 + 2 \,.\, D^2)}{\text{cone}}$.

23. Midddle Frustum of Sphere $= \frac{h(D^2 - \frac{1}{3}\text{ of } h^2)}{\text{cylinder}}$,

or $\frac{h(d^2 + \frac{2}{3}\text{ of } h^2)}{\text{cylinder}}$.

24. Any Frustum of Sphere

$= \frac{\frac{1}{2}\,h(d^2 + D^2 + 1\frac{1}{3}\text{ of } h^2)}{\text{cylinder}}$, or $\frac{2h(r^2 + R^2 + \frac{1}{3}\text{ of } h^2)}{\text{cylinder}}$.

25. Segment of Sphere $= \frac{h(h^2 = 3 \,.\, R^2)}{\text{globe}}$.

On examining the above it will be seen that the formulæ for frustums readily resolve themselves into those for their corresponding complete solids; thus, if the frustum in formula 16 is supposed to be completed, and run up to a point, then s vanishes, and the rule becomes $\frac{l\,(S^2 + S^2)}{\text{frustum}}$; and since the frustum divisor is double the pyramidical, the numerator and denominator cancel by 2, and become $\frac{l\,S^2}{\text{pyramid}}$. In the 18th, if d and r vanish, the formula is resolved into the 12th, and so of the rest. A pyramid, as before remarked, is equal to $\frac{1}{3}$, a parabolic prism to $\frac{2}{3}$ of its circumscribing rectangular prism. A cone is $\frac{1}{3}$, a sphere or spheroid $\frac{2}{3}$, a paraboloid $\frac{1}{2}$, and a parabolic spindle $\frac{8}{15}$, of its circumscribing cylinder. An examination of the formulæ for these solids will show that they are so constructed. Thus, comparing the 9th with the 8th, we find $2l$ instead of l; and comparing the 12th with the 3d, we have $\frac{1}{2}h$ for h or l. The parabolic spindle being $\frac{8}{15}$ of the cylinder, and the cylinder $\frac{3}{2}$ of the globe, multiplying these together we have .8. The difference between an oblate and prolate spheroid will be best understood by considering the revolutions of a parallelogram. Suppose a parallelogram 12 inches by 6 to be divided by two lines across the middle, at right angles to each other, so as to cut it into 4 equal portions, each 6 by 3. Then, if the parallelogram revolve on the short axis, it will generate a cylinder 6 inches deep, and having a diameter of 12 inches; consequently, its content will be $6 \times 12^2 \times .7854$. If it revolve on the long axis the cylinder produced will be 12 inches deep, and 6 diameter, and its content $12 \times 6^2 \times .7854$; that is, in each case $fv^2\,.7854$.* Two-

* In short, all the formulæ for round solids are but modifications of the general expression $\frac{fv^2}{G}$; and even angular solids

thirds of this, or fv^2 .5236, gives the spheroid $= \frac{fv^2}{\text{globe}}$. The earth is a spheroid slightly oblate, the polar diameter, as determined by careful measurements of a degree at different parts of its surface, being about 26 miles less than the equatorial, the prominence of the torrid zone having, it is presumed, been acquired, at the commencement, from the operation of centrifugal force: it being supposed that the earth was formed from matter in a semi-fluid state, and set rotating on its axis before the parts had been allowed time to consolidate.

ILLUSTRATION OF FORMULÆ.

74. Formula 1.—What is the content, in gallons, of a vessel in the shape of a square prism, 1 inch deep, and 29 inches along each of the sides?

Referring to the back of the rule, 16.65 will be found the gauge point for square prisms; therefore over 16.65 of D set 1; over 29 is 3.03 gallons, the content.*

75. Formula 2.—Each side of an hexagonal pyramid is 46 inches, its perpendicular depth 90 inches; what is the content in gallons? Over 17.9 set 90; over 46 is 594.8 gallons.

76. Formula 3.—The depth of a cylinder is 40 inches,

---

may come under the same form, if we conceive them to be described by the rotation of planes, and the generated surfaces subsequently shaped into polygons by lateral compression.

* Finding the content of solids whose depth or thickness is unity is generally termed "gauging areas," because, in such cases, the surface and solidity are both represented by the same number, ls^2 and s^2 being equivalent when l becomes 1.

its diameter 21.5; how many gallons will it contain? Over 18.78 set 40; over 21.5 is 52.37 gallons.

77. Formula 4.—The depth of a cone is 24 inches, its diameter 17; how many lbs. of tallow will it hold? Over 10.75 set 24; over 17 is 60 lbs.

78. Formula 5.—What is the weight of a globe of brass 8 inches in diameter? Over 2.51 set 8; over 8 is 81.2 lbs.

79. Formula 6.—The fixed, or transverse axis, of a prolate spheroid is 54 inches, its conjugate 33; how many bushels will it contain? Over 65.08 set 54; over 33 is 13.88 bushels.

80. The fixed, or conjugate diameter of an oblate spheroid is 33 inches, its transverse 54; how many bushels will it contain? Over 65.08 set 33; over 54 is 22.7 bushels.

81. Formula 7.—A cistern in the shape of a rectangular prism, or parallelopiped, is 82 inches long, 54 broad, and 37.5 deep; how many gallons will it contain? To 54 on D set 54 of the slide, then under 37.5 is 45, a mean proportional. Over 16.65 set 82; over 45 is 598½ gallons.

82. Formula 8.—A vessel oblong at top, and tapering downward to a point, measures 48 inches by 75; its depth is 63 inches; how many lbs. of hot hard soap will it hold? Over 48 of D set 48, then under 75 is 60, a mean proportional. Over 9.16 set 63; over 60 is 2700 lbs.

83. Formula 9.—A prismatic vessel, 10 inches deep, whose ends are in the shape of a parabola, measures 80

inches along the straight side or double ordinate, from the middle of which to the vertex is 60 inches; how many gallons will it contain? Over 60 set 60, under 80 is 69.28, a mean. Over 28.84 set 20 (twice depth;) over 69.28 is 115.4 gallons.

84. Formula 10.—The axes of an elliptic cylinder are 67 and 52, its depth 50 inches; how many bushels will it contain? Over 52 set 52; under 67 is 59, a mean. Over 53.14 set 50; over 59 is 61.6 bushels.

85. Formula 11.—The axes of an inverted elliptic cone are 16 and 9, the depth 19 inches; how many pints will it hold? Over 16 set 16; under 9 is 12, a mean. Over 11.5 set 19; over 12 is 20.6 pints.

86. Formula 12.—A vessel in the shape of a parabolic conoid is 42 inches deep, and the diameter of the top is 24 inches; what is the content in gallons? Over 18.78 set 21 (half 42;) over 24 is 34.25 gallons: or by the second, over 18.78 set 84 (twice 42); then over 12 is 34.25, as before.

87. Formula 13.—What is the content, in gallons, of a hyperbolic conoid, the diameter at top being 52 inches, the diameter in the middle 34, and the depth 25 inches? Over 46 set 25, then—

Over 52 is 31.9
Over 68 is 54.6

86.5 gallons.

88. Formula 14.—What is the content, in gallons, of a parabolic spindle, the diameter of which is 28 inches,

and length 70 inches? Over 23 set 70; over 28 is 103.7. Then to 103.7 on A set commencement of slide, and over .8 is 82.96 gallons.

89. Formula 15.—The length of a spindle is 20 inches, the greatest diameter 6 inches, and the diameter halfway between it and the point 4.74 inches; what is the content in cubic inches? Over 2.76 set 20, then—

Over 6.	is	94.
Over 9.48	is	235.5
		329.5 cubic inches.

90. Formula 16.—How many gallons will be contained in the frustum of an octagonal pyramid, each side of the greater base being 17.5 inches, of the less 14 inches, and the perpendicular depth 47 inches? Over 18.55 set 47, then—

Over 14.	is	26.8
Over 17.5	is	41.8
Over 31.5	is	135.2
		203.8 gallons.

91. How many lbs. of hot hard soap will the above contain?

As polygonal pyramids are not figures of frequent occurrence, it was not deemed necessary to insert gauge points for any other quantities than gallons and cubic feet, the weight therefore must be determined by a second process, which, since a gallon of water weighs 10 lbs., is effected by multiplying the content in gallons by 10 times the specific gravity. Now the specific gravity of hot hard soap is shown on the rule to be .99, ten times which = 9.9. Therefore, to 203.8 on A set commencement of slide, then over 9.9 is 2017 lbs.

92. Formula 17.—The frustum of a cone is 43 inches deep, the diameter at one end 36, at the other 20 inches; how many bushels will it contain? Over 130.17 set 43, then—

Over 20 is 1.02
Over 36 is 3.28
Over 56 is 7.96

12.26 bushels.*

93. Formula 18.—The diameters of the frustum of a paraboloid are 30 and 40 inches, the depth 18 inches; how many gallons will it contain? Over 18.78 set 9 (half 18,) then—

Over 30 is 23.
Over 40 is 40.7

63.7 gallons.†

94. Formula 19.—How many bushels will be contained in the frustum of a hyperbolic conoid, the top and bottom

* If the frustums of two equal cones be joined together at their greater ends they form a figure called by gaugers a cask of the 4th variety.

† If the frustums of two equal paraboloids be joined together at their greater ends they form a figure called by gaugers a cask of the 3d variety.

diameters of which are 23 and 40 inches, the middle 36 inches, and depth 20 inches? Over 130.17 set 20, then—

Over 23 is .62
Over 40 is 1.88
Over 72 is 6.13

8.63 bushels.

95. Formula 20.—The length of a vessel in the form of the middle frustum of a parabolic spindle is 20, the greatest diameter 16, and least 12 inches; what is the content in gallons? Here twice the difference of the diameters = 8; therefore, over 32.54 set 20, then—

Over 12 is 2.72
Over 16 is 4.83
4.83

12.38
Over 8 is 1.2, one-tenth of which is .12

12.26 gallons.*

96. Formula 21.—The bung diameter of a vessel is 36 inches, the head 30, twice the diameter taken midway between them 67.8 inches, and the length 40 inches; how many gallons will it contain? Over 46 set 40, then—

Over 30 is 17.
Over 36 is 24.5
Over 67.8 is 86.9

128.4 gallons.

* A cask in the form of the middle frustum of a parabolic spindle is termed by gaugers a cask of the 2d variety.

97. Formula 22.—A vessel in the form of the middle frustum of a prolate spheroid is 40 inches long, the bung diameter is 36, and the head 27 inches; what is the content in gallons? Over 32.54 set 40, then—

Over 27 is 27.4
Over 36 is 49.
49.

125.4 gallons.*

98. Formula 23.—How many cubic feet are contained in the middle zone of a sphere, the axis of which is 44 inches, and the height of the zone 14 inches? Over 46.9 set 14, then—

Over 44 is 12.32
Over 14 is 1.26, one-third of which = .42

11.9 bushels.

99. Formula 24.—What is the content in gallons of the shoulder of a still in the form of the frustum of a sphere, the top and bottom diameters being 42 and 36 inches, and the height 30 inches? Over 18.78 set 15 (half 30,) then—

Over 36 is 55
Over 42 is 75
Over 30 is 38.3
+ $\frac{1}{3}$ = 12.7

181. gallons.

* A cask in the form of the middle frustum of a prolate spheroid is termed by gaugers a cask of the 1st variety.

100. Formula 25.—A copper basin in the form of the segment of a sphere is 18 inches deep, the diameter across the top 40 inches; how many gallons will it contain? Over 23 set 18, then—

Over 18 is 11.
Over 20 is 13.6
+ twice ditto = 27.2

51.8 gallons.

The content of cylindroids, prismoids, and wedges, is found by taking the mean proportionals of the products of the top and bottom dimensions, and of the product of their sums, making use of the round frustum gauge points for the cylindroid, and the square frustum for the prismoid and wedge.

ILLUSTRATION.

101. The perpendicular depth of a cylindroid is 52 inches, the diameters at top 60 and 46, at bottom 42 inches; what is the content in bushels?

Bottom	42 × 42,	mean proportional between which		=42
Top	60 × 46	"	"	=52.54
Sum	102 × 88	"	"	*=94.74

Over 130.17 set 52, then over 42 = 5.4
52.54 = 8.5
94.74 = 27.5

41.4 bushels.

Referring to the Table on page 150, the round frustum divisor for bushels is 16945.74.

The numerical solution of this question, therefore, will be as follows:—

```
Bottom  42 × 42 = 1764
Top     60 × 46 = 2760
Sum    102 × 88 = 8976
                 -----
                 13500
                    52
                 -----
                 27000
                67500
                --------
16945.74 ) 702000.00 ( 41.426 bushels.
           6778296
           -------
            2417040
            1694574
            -------
             7224660
             6778296
             -------
              4463640
              3389148
              --------
              10744920
              10167444
              --------
                577476
```

For questions 102 and 103 the divisor, as shown on page 150, will be 13309.15.

102. The length and breadth of a coal wagon at top are 81 and 55 inches, at bottom 41 and 29 inches; the depth is 47 inches; how many bushels will it contain?

Top	81×55, mean proportional between which			=66.8
Bottom	41×29	"	"	= 35.5
Sum	122×84	"	"	=101.2*

---

* It must be observed that, in irregular solids, the mean proportional of the sum is not the sum of the mean proportionals;

Over 115.3 set 47, then over 66.8 = 15.6
35.5 = 4.4
101.2 = 36.1

56.1 bushels.

103. A heap of malt is piled into the form of a wedge, 24 inches deep, the base is 40 inches long, and 20 broad, the edge 20 inches long; how many bushels does it contain?

Top 20 × 0
Bottom 30 × 20, mean proportional between which = 28.28
Sum 60 × 20, " " = 34.37

Over 115.3 set 24, then over 28.28 = 1.46
34.36 = 2.14

3.6 bushels.

---

the former must be taken, not the latter. In prismoids, also, attention must be paid to the *position* of the sides; for the top and bottom areas of two prismoids may be the same, and yet their middle area, and consequently their content, different. For, suppose a prismoid to be 12 inches by 10 at top, and 9 inches by 6 at bottom, the 12 falling over the 9, and the 10 over the 6: if now we shift the position of the top parallelogram so as to bring the short side over the long one of the bottom, then the figure becomes distorted, and the content altogether altered.

In the first it will be—		In the second—	
Bottom	12 × 10 = 120	Bottom	12 × 10 = 120
Top	9 × 6 = 54	Top	6 × 9 = 54
Sum	21 × 16 = 336	Sum	18 × 19 = 342
	$510 \times \frac{1}{6} l.$		$516 \times \frac{1}{6} l.$

The following Table of Divisors will enable the Student to solve the preceding Questions numerically. The same formulæ apply.

DIMENSIONS IN INCHES.	Specific gravity.	SQUARE SOLIDS.			ROUND SOLIDS.			
Content in		Prism.	Pyramid.	Frustum.	Cylinder.	Globe.	Cone.	Frustum.
Cubic Inches		1.	3.	6.	1.2732	1.90985	3.8197	7.6394
Cubic Feet		1728.	5184.	10368.	2200.16	3300.24	6600.48	13200.96
Pints		34.659	103.978	207.956	44.129	66.194	132.388	264.776
Gallons		277.274	831.82	1663.64	353.036	529.554	1059.108	2118.216
Bushels		2218.19	6654.57	13309.15	2824.29	4236.435	8472.87	16945.74
Hot hard Soap, lbs.	.99	28.	84.	168.	35.65	53.475	106.95	213.9
Cold ditto	1.02	27.14	81.42	162.84	34.55	51.832	103.664	207.327
Tallow	.915	30.28	90.84	181.68	38.55	57.83	115.65	231.3
Flint Glass	3.21	8.64	25.92	51.84	11.	16.5	33.	66.
Plate ditto	2.418	11.3	33.9	67.8	14.4	21.6	43.2	86.4
Platinum	21.45	1.29	3.87	7.74	1.642	2.463	4.927	9.854
Gold	19.25	1.44	4.32	8.64	1.8	2.75	5.5	11.
Mercury	13.61	2.03	6.09	12.18	2.584	3.877	7.753	15.5
Lead	11.35	2.44	7.32	14.64	3.11	4.67	9.34	18.68
Silver	10.53	2.64	7.92	15.84	3.36	5.04	10.08	20.16
Copper	8.81	3.155	9.46	18.92	4.	6.	12.	24.
Brass	8.41	3.3	9.91	19.82	4.2	6.3	12.6	25.2
Wt. Iron & Steel	7.82	3.54	10.62	21.24	4.5	6.75	13.5	27.
Ct. Iron, Tin, & Zinc	7.24	3.8	11.41	22.82	4.8	7.2	14.44	28.8
Ice & Gunpowd.	.93	29.81	89.43	179.86	37.94	56.91	113.82	227.64

POLYGONAL SOLIDS.

DIMENSIONS IN INCHES.	CONTENT IN GALLONS.			CONTENT IN CUBIC FEET.		
	Prism.	Pyramid.	Frustum.	Prism.	Pyramid.	Frustum.
Trigonal	640.34	1921.01	3842.02	3090.64	11971.93	23943.87
Tetragonal	277.274	831.82	1663.64	1728.	5184.	10368.
Pentagonal	161.161	483.48	966.96	1004.37	3013.12	6026.23
Hexagonal	106.72	320.16	640.32	665.11	1995.32	3990.64
Heptagonal	76.39	229.17	458.34	475.52	1426.56	2853.12
Octagonal	57.42	172.27	344.54	357.88	1073.64	2147.28
Nonagonal	44.85	134.56	269.12	279.53	838.58	1677.17
Decagonal	36.03	108.11	216.22	224.58	673 75	1347.50
Undecagonal	29.6	88.81	177.63	184.5	553.51	1107.02
Dodecagonal	24.76	74.30	148.6	154.34	463.02	926.03

Numerical solution of Question 78. $\frac{8^3}{6.3} = 81.2$ lbs.

Numerical solution of Question 90.

Top $14.^2 = 196$
Bottom $17.5^2 = 306.25$
Sum $31.5^2 = 992.25$

$1494.5 \times 47 = 70241.5$; and $\frac{70241.5}{344.54} = 203.8$ gallons.

EXAMPLES FOR PRACTICE.

104. What is the weight of a prism of steel 7 inches square, 15 inches long?—Ans. 207 lbs.

105. What would be the weight of a pyramid of ice, 8 inches square at bottom, and 13.8 inches high?—Ans. 9.87 lbs.

106. A cylindrical glass-pot, 24 inches diameter, is charged with flint glass to the depth of 15 inches: what is its weight?—Ans. 785 lbs.

107. An inverted cone is 23 inches deep, its diameter at top 10 inches: what quantity of tallow will it contain? —Ans. 19.8 lbs.

108. What quantity of gunpowder, shaken down, will fill a shell whose internal diameter is 9 inches?—Ans. 12.8 lbs.

109. The axes of an oblong or prolate spheroid are 6 and 8 inches: what quantity of mercury will it contain? —Ans. 74.2 lbs.

110. The axes of an oblate spheroid are 6 and 8 inches: what quantity of mercury will it contain?—Ans. 99 lbs.

111. What is the weight of a rectangular block of ice, 12 inches by 10 thick, and 30 inches long?—Ans. 121 lbs.

112. The top of an inverted rectangular pyramid measures 17 inches by 13; its depth is 44 inches: how many gallons of water will it contain, and how many lbs.?—Ans. 11.69 gallons, 116.9 lbs.

113. The base of a parabola is 32 inches, its absciss 24 inches; the depth is 1 inch: how many gallons will it hold?—Ans. 1.84 gallons.

114. The diameters of an elliptic cylinder are 25 and 20 inches, the depth 13 inches: how many gallons will it contain?—Ans. 18.4 gallons.

115. An elliptic cone of silver is 10 inches high, the diameters at bottom 5 inches by 4: what is its weight in lbs. avoirdupois?—Ans. 19.8 lbs.

116. A paraboloid of copper is 12 inches high, the diameter of the base 8 inches: what is its weight?—Ans. 96 lbs.

117. A vessel in the shape of an hyperboloid is 25 inches deep, the radius of the top 26, and the middle diameter 34 inches: what quantity of cold hard soap will it hold?—Ans. 883 lbs.

118. The length of a parabolic spindle is 32 inches, its diameter 10 inches: required the content in gallons.—Ans. 4.83 gallons.

119. The length of a cast-iron spindle is 20 inches, its greatest diameter 9 inches, and the diameter halfway between that and the point 6 inches: what is its weight?—Ans. 155.8 lbs.

120. The frustum of a nonagonal pyramid, 25 inches deep, measures 9 inches along each side at top, and 12 at bottom: how many gallons will it contain?—Ans. 61.8 gallons.

121. Suppose a cask to consist of two equal frustums of a cone, the length of which is 40 inches, the bung dia-

meter 32, and the head 24 : what is the content in gallons? By formula 17.—Ans. 89.4 gallons, 4th variety.

122. Suppose a cask, of the same dimensions, to be composed of two equal frustums of a paraboloid: required the content. By formula 18.—Ans. 90.6 gallons, 3d variety.

123. Let the cask be the middle frustum of a parabolic spindle, and the dimensions remain the same: what is the content? By formula 20.—Ans. 98.2 gallons, 2d variety.

124. Let the cask be the middle frustum of a prolate spheroid, the dimensions continuing the same: what is the content? By formula 22.—Ans. 99.1 gallons, 1st variety.

125. What will be the content of the middle frustum of a spindle having the same dimensions, and also the diameter halfway between the head and bung 29.6 inches? By formula 21.—Ans. 96.45 gallons, true content.

126. Required the content in cubic feet of the middle frustum of a sphere, the height of which is 24, and the least diameter 18 inches.—Ans. 7.72 feet.

127. Find the content in gallons of the frustum of a sphere, the height of which is 9 inches, and the radii at its ends 14 and 10 inches.—Ans. 16.47 gallons.

128. What is the weight of the segment of a globe of lead, the height of which is 6 inches, and the radius of the base 8 inches?—Ans. 293 lbs.

129. The depth of a cylindroid is 50 inches, the diame-

ters of the elliptic base are 60 and 44 inches, the diameter of the circular top 40 inches: required the content in gallons.—Ans. 298.3 gallons.

130. The depth of a prismoid is 50 inches; the base is a parallelogram 60 inches long, 44 broad; the top is a square, the sides of which are each 40 inches: what is the content in gallons?—Ans. 379.8 gallons.

131. The frustum of a square pyramid is 30 inches deep, each of the sides at bottom 36, and at top 25 inches: what is the content of each of the wedges into which a diagonal plane, passing through its extremities, divides it?—Ans. 62.97 gallons, lower hoof or wedge; 38.77 ditto, upper ditto.

TABLES VII. and VIII., at the back of the rule, are adapted for the use of the E slide. Table VII. exhibits the weight of metallic spheres, commencing thus:—

	D diameter.	E weight.
Platinum	4 inches =	26 lbs. avoirdupois.
Gold	6½ inches =	100 lbs.

That is, over 4 on D, set 26 lbs. on E, then the numbers on D will be a series of diameters, and the numbers over them on E their corresponding weights.

EXAMPLES.

132. A sphere of platinum weighs 51 lbs.: what is its diameter?—Over 4 set 26 lbs.; under 51 lbs. is 5 inches.

133. A sphere of silver weighs 7 lbs.: what is its diameter?—Ans. 3.284 inches.

134. Rockets receive their names from a comparison of the external diameters of their cases with leaden balls: what, then, is the diameter of a 5-pound rocket?—Ans. 2.86 inches.

135. A globe of wrought iron weighs 19.7 lbs.: what is its diameter?—Ans. 5.1 inches.

136. A spherical vessel, filled with mercury, holds 258 lbs.: what is its diameter?—Ans. 10 inches.

137. Thirteen lbs. of gunpowder fill a shell: what is its diameter?—Ans. 9.04 inches.

138. A sphere of brass weighs 81.2 lbs.: what is its diameter?—Ans. 8 inches.

Table VIII. is used precisely like Table VII., and is for finding the diameters and circumferences of spheres from their solidities; and also the solidities of regular bodies, the tetrahedron, &c.

EXAMPLES.

139. The solidity of a sphere is 33.6: what is its diameter?—Over 4.6 of D set 51 of E; under 33.6 is 4.

140. A globe contains 98.5 solid feet: what is its circumference?—Ans. 18 feet.

141. The side of a tetrahedron measures 2.2 inches: how many cubic inches does it contain?—Ans. 1.25 cubic inches.

142. The side of an octahedron measures 3.3 inches: how many cubic inches does it contain?—Ans. 16.875 solid inches.

143. A dodecahedron contains 15 cubic feet: what is the length of each of its sides?—Ans. 1.25 feet.

144. The solidity of an icosahedron is 162: what is the length of each of its sides?—Ans. 4.2.

SOLAR SYSTEM.

THE concluding part of the tabular work on the rule is for the use of the line A in conjunction with that of E. According to Kepler's famous discovery, the squares of the periodic times of the planets are proportional to the cubes of their mean distances. Now, since the line A is laid down twice, and the line E thrice, in the same space, when the slide E is laid evenly in, the cubes of the numbers on A will be equal to the squares of the numbers on E; when in any other position, the cubes of the numbers on A will be *proportional* to the squares of the numbers on E. Hence, if under 95 millions of miles on A we set 365 days, or 52 weeks, or 13 lunar months, or 1 year, on E; then the numbers on A will be a series of planetary distances, and the numbers beneath them on E their periods of revolution, in days, weeks, months, or years, according as 365, 52, 13, or 1, is selected.

EXAMPLES.

145. The distance of Mercury from the sun is 37 millions of miles; what is the length of his year?—Under 95 set 365; under 37 is 88 days.

146. Mars is about 687 days in revolving round the sun; what is his distance?—Ans. 144 millions.

147. Herschel's mean distance is about 1823 millions of miles; how many years does he consume in traversing his orbit?—Ans. 83.8 years.

148. If a planet revolved in an orbit 20 million miles from the sun; how long would it take in passing round him?—Ans. 35¼ days.

149. Suppose the recently discovered planet to be 3,000 millions of miles distant from the sun; how many years does it take in traversing its orbit?—Ans. 178 years.

150. The nearest of Saturn's moons is 108 thousand miles distant from him, and the time of its periodic revolution about 22¾ hours; the second is distant 140 thousand: what is its periodic revolution?—Under 108 of A set 22¾ of E; under 140 is 34 hours nearly.

151. The fourth satellite of Saturn spends 65 hours in passing round its primary; required its distance from him.—Ans. 217,000 miles.

The following table will enable the student to solve the previous questions numerically:—

WEIGHT OF METALLIC SPHERES.

Platinum	4	inches diam.	=	26 lbs. av.
Gold	6.5	...	=	100
Mercury	3	...	=	7
Lead	7.5	...	=	90
Silver	4.5	...	=	18
Copper	6	...	=	36
Brass	5.4	...	=	25
Wt. Iron and Steel	3	...	=	4
Ct. Iron, Tin, and Zinc	6	...	=	30
Ice and Gunpowder	7	...	=	6

Solidity of Sphere $= .5236\ d^3 = .01688\ c^3$.

Solidity of Tetrahedron $= .11785\ s^3$.
" Octahedron $= .47140\ s^3$.
" Dodecahedron $= 7.66312\ s^3$.
" Icosahedron $= 2.18169\ s^3$.

Numerical Solution of Question 133.

$$\overset{\text{lb.}}{18} : \overset{3}{4.5} :: \overset{\text{lb.}}{7} : d^3;$$ or taking the $\sqrt[3]{}$ of each term

$\sqrt[3]{18} : 4.5 :: \sqrt[3]{7} : \frac{4.5\sqrt[3]{7}}{\sqrt[3]{18}}$; which, multiplying numerator and denominator by $\sqrt[3]{18^2} = \frac{4.5\sqrt[3]{2268}}{\sqrt[3]{18^3}} = \frac{1}{4}\sqrt[3]{2268} = 3.284$.

Question 148.

$$95^3 : 365^2 :: 20^3 : \frac{365^2 \times 20^3}{95^3} = \frac{133225 \times 8000}{857375} = 1243;$$

and $\sqrt{1243} = 35.25$ days.

MISCELLANEOUS QUESTIONS.

152. A triangular piece of board, measuring 18 feet in perpendicular height, is to be divided equally among 4 men, by sections parallel to the base: at what distance from the vertex must they be cut?—Similar surfaces vary as their squares; hence, over 18 of D, set 4 shares on C; then under 3 shares is 15.58 feet; under 2 is 12.72, and under 1 is 9 feet.

153. A circle measures 9 inches in diameter: required the diameter of another of twice the area.—Over 9 of D, set 1; under 2 is 12.72.

154. Four men bought a grindstone, 30 inches in diameter, and agreed that the first should use it till he ground down $\frac{1}{4}$ of it for his share, deducting 6 inches in the middle for waste; then, that the second should use it till he

ground down $\frac{1}{4}$ part, and so on: what part of the diameter must each grind down?—If $\frac{1}{5}$th of the diameter be waste, $\frac{1}{25}$th of the content is waste; therefore, conceiving the whole to contain 25 shares, 1 share will be waste, and each man will have 6 shares. Over 6 inches on D, set 1 share on C; under 7 is 15.87; under 13 is 21.63; under 19 is 26.15; and under 25 is 30. Subtracting these numbers from each succeeding one, we obtain 9.87 inches for the fourth; 5.76 for the third; 4.52 for the second; and 3.85 for the first.

155. Three persons having bought a sugar-loaf, 20 inches high, it is required to divide it equally among them by sections parallel to the base: required the height of each part.—Similar solids vary as their cubes, hence use the E slide. Over 20 of D, set 3 shares; under 2 is 17.48; under 1 is 13.86. Subtracting from each preceding, we have 2.52 inches height of lowest part: 3.62 second; 13.86 third.

156. A person has a solid globe of wood, 7 inches in diameter, and requires another twice the size: required its diameter.—Over 7 of D, set 1 of E; under 2 is 8.82 inches.

157. Perceiving a chandelier, suspended from a church ceiling, moving slowly backwards and forwards, I observed that it made 14 swings per minute: what was the height of the ceiling from the floor, supposing the centre of gravity of the chandelier to be 8 feet from the pavement?—Ans. 68 feet.

158. A person lent another a cubical rick of hay, measuring 10 feet each way, which he repaid with 3 others

of the same shape: what was the measure of each?—Ans. 6.93 feet.

159. Two pipes, each 2 inches internal diameter, fill a cistern in an hour; they are then stopped, and five smaller ones are opened at the bottom of the vessel, which they empty in the same space of time: what is the diameter of these smaller pipes, each being the same?—Ans. 1.265 inches.

160. The arms of a pair of scales are of unequal length; a quantity of sugar, weighing 19 lbs. in one scale, weighs only 16 lbs. in the other: what is its real weight? Take the mean proportional.—Ans. 17.43 lbs.

161. There is a glass in the shape of a frustum of a cone, 6 inches deep; its top diameter is 3 inches, its bottom 2; if I pour water into it till it is $\frac{4}{5}$ full, what will be the depth of the liquor?—Ans. 5.12 inches.

162. Three men bought a tapering piece of timber, which was the frustum of a square pyramid: each side of the greater end was 3 feet, of the less 1 foot, the length was 18 feet: what was the thickness of each man's piece, supposing they are to have equal shares?—Ans. 3.27, 4.56, and 10.17 inches.

163. The sides of a triangle measure 6, 5, and 3 feet; it is required to construct another that shall contain $3\frac{4}{5}$ times as much; determine the length of the sides.—Ans. 11.69, 9.75, and 5.85.

164. The mean distance of Jupiter from the sun is 495 millions of miles: how many years is this splendid luminary in traversing his orbit?—Ans. $11\frac{7}{8}$ years.

165. How many gallons will be contained in a cylin-

drical vessel, 18½ inches diameter, and 8¼ deep?—Ans. 8 gallons.

166. A globe of cast-iron weighs 18 lbs.: what is its diameter?—Ans. 5.09 inches.

167. What is the diameter of a silver sphere, weighing 25 lbs. avoirdupois?—Ans. 5.02 inches.

168. Seven men bought a grindstone, a yard in diameter, for a guinea; they paid 3*s.* each, and agreed to grind down their separate portions in succession: what was the diameter of the stone when each began to grind?—Ans. 36, 33.3, 30.4, 27.2, 23.57, 19.24, and 13.6 inches.

169. There are two similar cylinders; the length of the one is 8 inches, and its diameter 4; the other is $2\frac{4}{5}$ times the size: determine its length and diameter.—Ans. Length 11.28, diameter 5.64.

170. Two spheres of brass are to each other in the proportion of 5 to 7; if the larger measures 12 inches round, what is the circumference of the smaller?—Ans. 10.72.

171. Five men bought a grindstone 16 inches in diameter, for 15*s.* A pays 1*s.*, B 2*s.*, C 3*s.*, D 4*s.*, and E 5*s.*; each man is to grind down his portion in succession, commencing with A, and ending with E, who is to leave 4 inches unground: what is the diameter as each begins to grind?—Ans. 16, 15.49, 14.42, 12.65, 9.8.

172. The frustum of a pentagonal pyramid measures 10 inches along each side at top, and 15 at bottom, and the depth is 20 inches; if I put into it a solid globe of wrought iron, weighing 108 lbs., and then pour in 12 gallons of water: what depth of the vessel will remain unfilled?—Ans. 8.28 inches.

173. A globe of wood, 10 inches diameter, suspended in a lathe, was turned down into smaller globes, by 4 men successively, each chipping off an equal portion: what was the diameter when each began, supposing the last left a globe of 2 inches diameter?—Ans. 10, 9.09, 7.96, and 6.35.

174. If into a soap-bubble $3\frac{3}{5}$ inches diameter, I blow $\frac{7}{9}$ of a pint of air, what will then be its circumference?—Ans. 14.49 inches.

175. One evening I chanced with a tinker to sit,
Whose tongue ran a great deal too fast for his wit.
He talked of his art with abundance of mettle,
So I asked him to make me a flat-bottom'd kettle.
Let the top and the bottom diameters be
In just such proportion as five are to three.
Twelve inches the depth I proposed, and no more;
And of gallons to hold seven-tenths of a score.
He promised to do it, and straight to work went,
Got right the proportions, but wrong the content.
He alter'd it then, and the quantity found
Correct, but the top measured far too much round;
Till, making it either too big, or too little,
The tinker, at last, had quite spoil'd his fine kettle.
But he vows he will bring his said promise to pass,
Or else that he'll waste every ounce of his brass.
So to save him from ruin, kind friend, find him out
The diameters' length, for he'll ne'er do it, I doubt.
Ans. 15.06 bottom, 22.1 top.

CASK GAUGING.

It has been stated, that casks are usually gauged by considering them under four varieties; and questions 121, 122, 123, 124, show the content of a cask of given dimensions under these four varieties, in which it will be seen that there is a variation of 10 gallons, according to the different form under which the cask is viewed. By considering the vessel as part of a prolate spheroid, we shall have the content too great; for no cask is so much curved towards the head as this would make it. The middle frustum of a parabolic spindle approaches nearer to the shape of casks in general. Two frustums of a paraboloid leave too sharp a ridge in the middle; and the frustums of two cones give the content far too small, and would, in themselves, make a ridiculous kind of barrel. The generality of casks seem to be a compound of the first and fourth varieties, the bung part being spheroidal, and the extremities conical. If, in addition to the bung and head, we take the diameter halfway between the two, then the true content is readily found by the general rule for frustums; but, as in practice, except with open casks, it is a somewhat tedious process to obtain this middle diameter *perfectly correct*, without which it is useless, since, by the nature of the formula, the content is made to depend upon it in a *fourfold* measure; and as the determining to which of the varieties any given cask makes the nearest approach, is a work requiring much skill and judgment, various writers have, from time to time, attempted to discover a rule that shall be an approximation for all casks. Dr. Hutton's, for this purpose, is represented by the formula $l\,(39\ B^2 + 25\ H^2 + 26\ HB)\ .000031473$. This, besides being very laborious, generally gives the content too small. By considering the bung diameter as the principal regulator, and by combining the formulæ for spheroidal and conic frustums, we shall arrive at a method which never *can be far* from the truth, and is of the easiest application possible, as it may be put under the following form, with a whole number for a gauge point, a desideratum in all cases with the slide-rule.

$$\frac{l\,(H^2 + 2.B^2)}{33^2};$$

that is, over 33 on D set the length; then the number over the head plus twice the number over the bung, is the content in gallons. For computation, the formula is equally simple and easy.

$$l\,(H^2 + 2.B^2)\ .000919.$$

EXAMPLE.

176. A cask measures 47 inches long; the head diameter is 26, and the bung 31 inches: required the content.

Over 33 set 47; then

over 26 = 29.2
31 = 41.5
41.5

112.2 gallons.

Numerically, $26^2 = 676$
$31^2 = 961$
961

$\underline{2598} \times 47 \times .000919 = 112.2$ gallons.

This is an example of a cask whose middle diameter was found to be 29.3 as nearly as possible, the content of which in gallons, by formula 21, will be

$$\frac{l\,(H^2 + B^2 + \overline{2m|}^2)}{46^2}$$

Over 46 place 47; then

over 26. = 15.
31. = 21.4
58.6 = 76.1

112.5

The same formula, arranged for numerical computation, is

$$l\,(H^2 + B^2 + \overline{2m|}^2)\ .0004721.*$$

Example, $26^2 = 676$
$31^2 = 961$
$58.6^2 = 3433.96$

$5070.96 \times 47 \times .0004721 = 112.51$ gallons, true content.

---

* .0004721 is the reciprocal of the round frustum divisor for gallons.

By Dr. Hutton's Rule, the content will be—

$$39 \times 31^2 = 37479$$
$$25 \times 26^2 = 16900$$
$$26 \times 26 \times 31 = 20956$$
$$75335 \times 47 \times .000031473 = 111.42 \text{ gallons.}$$

The following Table contains 50 casks that were carefully gauged while empty, and their contents subsequently tested by actual measurement with water. A great portion are taken from Dr. Hutton's works, some from Nesbit and Little's gauging, some from Todd's Manual, and the rest have come under my own observation. They will serve as exercises for the student, and show the value and efficiency of the rule I have proposed.

No.	L	H	B	2 m	True Content, by Rule for Frustums 4 dimensions.	Content by Dr. Hutton's Rule for 3 dimensions.	Content by proposed Rule for 3 dimensions.
1	28.3	23.2	27.7	52.6	54.41	53.51	53.91
2	29.8	22.2	26.	49.6	51.05	50.41	50.52
3	30.8	23.2	27.5	52.2	58.44	57.78	58.10
4	32.2	24.5	30.1	56.8	71.94	70.51	71.38
5	30.	24.7	29.2	55.2	63.87	63.40	63.80
6	32.5	23.8	28.2	53.6	64.97	64.12	64.51
7	34.3	26.3	33.5	62.2	92.02	90.72	92.35
8	34.5	26.4	33.	61.4	90.49	89.71	91.01
9	41.	26.3	32.2	60.4	104.07	102.41	104.10
10	37.	26.1	31.8	59.8	92.03	90.89	91.85
11	44.5	34.4	40.8	77.6	186.34	183.61	184.53
12	47.	26.3	33.8	62.8	128.2	125.82	128.2
13	34.2	27.2	33.8	62.8	94.07	93.21	94.8
14	47.	25.3	32.	59.4	115.21	113.92	116.
15	45.5	30.7	38.	71.	159.54	157.01	159.56
16	44.6	24.7	32.2	59.2	108.47	107.12	109.82
17	48.6	24.2	32.1	58.8	116.40	114.78	117.61
18	46.	25.7	34.7	63.4	127.78	125.11	129.32
19	48.8	24.2	32.1	58.8	116.88	115.28	118.21
20	51.2	23.3	31.	56.4	113.24	112.53	115.98*
21	48.	28.2	33.8	62.8	133.28	132.03	135.87*
22	51.6	36.6	41.6	79.2	227.59	228.52	227.62
23	57.	32.7	42.	78.2	240.8	235.43	240.81
24	54.	34.8	44.8	83.	257.66	253.41	259.22
25	45.6	28.	34.6	64.8	133.04	131.37	133.05
26	45.	27.	36.	66.	135.56	133.25	137.12
27	46.7	24.6	30.9	58.	108.56	106.00	107.83
28	32.5	21.4	26.2	49.6	55.31	53.87	54.65
29	27.7	19.6	23.4	44.2	37.73	37.33	37.65
30	42.	28.	35.	65.4	124.64	122.81	124.62
31	50.	26.	32.	60.2	125.67	123.16	125.2
32	66.	52.	62.	116.6	627.63	627.21	630.31*
33	60.	36.	45.	84.	293.93	290.11	294.72
34	58.5	40.1	49.8	93.	351.76	349.42	352.10
35	60.	40.	50.	93.2	362.17	358.15	363.75
36	44.5	34.4	40.8	77.2	185.04	183.52	184.54
37	35.5	30.	36.	68.	114.3	112.97	113.93
38	49.5	23.3	31.7	58.6	116.41	112.18	116.13
39	40.	30.	36.	67.8	128.38	127.32	128.35
40	50.	23.	31.	57.4	112.94	108.78	112.62
41	54.	39.	45.	86.	278.95	276.31	276.46
42	51.	23.5	31.5	58.4	119.3	115.21	118.91
43	39.6	29.4	35.5	67.	123.64	121.92	123.28
44	39.5	30.	36.7	69.2	131.2	128.87	130.46
45	40.5	29.8	35.5	67.	126.9	126.06	126.86
46	41.5	30.	37.	69.6	139.36	136.72	138.75
47	51.	23.7	31.9	59.4	122.98	117.73	121.72
48	48.	25.2	33.4	61.8	126.22	122.78	126.4
49	56.4	32.3	41.2	76.4	230.63	225.21	230.03
50	45.3	28.2	34.9	65.2	134.12	132.55	134.52
				Total	7415.12	7312.05	7423.04

In 47 out of the 50, the rule I have proposed agrees nearest with the truth; in the three marked with an asterisk, Dr. Hutton's comes nearer. The total error by his is 103 gallons; by mine, 8 gallons. In setting down the dimensions, the most concise way will be to place the length on the left hand, with a brace between it and the diameters, recollecting that 33 is the gauge point, when three dimensions are used; and 46, when four dimensions are taken.

EXAMPLE.

177. The length of a cask is 40 inches, the head 30, the bung 36, and twice the middle 67.8 inches: required the content.

By Proposed Rule for 3 dimensions.

$$40\begin{cases}30 = 33.05\\ 36 = 47.6\\ \qquad 47.7^*\end{cases}$$

128.35 gallons.

By General Rule for 4 dimensions.

$$40\begin{cases}30 \quad = 17.\\ 36 \quad = 24.5\\ 67.8 = 86.9\end{cases}$$

128.4 gallons.

The dimensions of casks are taken most readily with the long and cross callipers, and the bung and head rods.

* If, as in this case, the number found standing over the bung diameter appear to be more than 47.6, and less than 47.7, set down both, as above.

When these are not at hand, their place may be supplied as follows:—

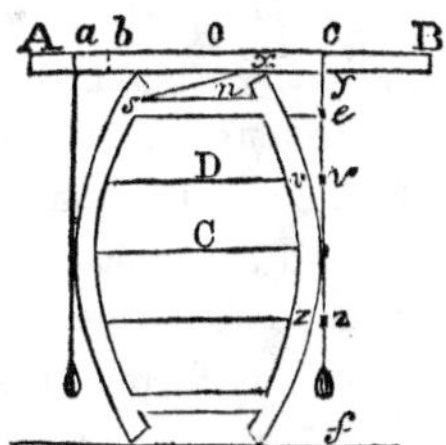

Procure a straight piece of deal, about $\frac{5}{8}$ of an inch square, and 6 feet long, for a measuring-rod; and, with a camel's-hair pencil and Indian ink, divide it into inches and tenths. Take another piece, AB, an inch square, and about 4 feet long; and near one end, as at *a*, cut a notch, and 2 inches from it, *b*, make a mark, and place a cipher 0. Then divide the distance from *b* to the end B into inches and tenths. Also procure two pieces of string, each with loops at one end, and heavy plummets of lead at the other. Before tying the loop, on one of the strings slip 3 pieces of cork, *e*, *v*, *z*, about $\frac{1}{8}$ of an inch thick, and $\frac{1}{4}$ of an inch square. Then,

To take the Dimensions of a standing Cask.

With a piece of string and chalk, by problem 2, page 20, strike a line across the middle of the head of the cask; lay the rod AB over this line, and bring the plummet depending from *a* up to the bulge of the cask. Then slip the other plummet along to *c*, till it touches the cask in like manner. The number now cut by *c* will be the internal bung diameter C, the distance *ab*, of 2 inches, being an allowance for twice the thickness of the staves. With the measuring rod take the distance from *y* (the *under* side of the rod AB) to the ground *f*. Also the distance from *o* (the upper side of the rod) to *n*, the head of the barrel. Then *yf* minus twice *on*, will be the internal length of the cask, the thickness of the square rod, AB, being supposed equal to the thickness of the head of the cask, which is generally 1 inch. To take the middle diameter D, slip the top cork up to *e*, till the distance *ye* is equal to *on*. The length of the cask being known, slip the second cork down to *v*, the distance *ev* being $\frac{1}{4}$ of the length; in the same manner adjust the cork *z*, if deemed necessary. Then add together the distances *vv*, *zz*, and subtract their sum from the bung diameter, or deduct twice the distance *vv*, if the curve of the cask be uniform; the remainder will be the middle diameter, D. In the same way might a diameter be taken halfway between D and C. The oblique line *sx*, measured from the inside of the chimb to the outermost sloped edge of the opposite stave, will be the internal head diameter; or twice the distance at *e* may be deducted from the bung. If only three dimensions are taken, the corks may be dispensed with; but in ullaging standing casks, they will be found extremely convenient.

For taking the dimensions of lying casks, a common pair of callipers may be made by any carpenter, as annexed.

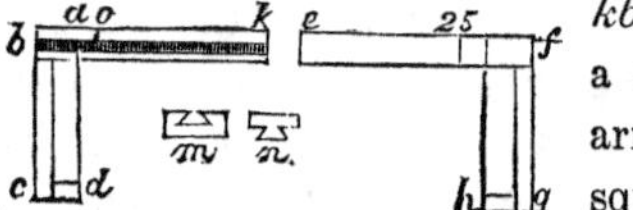

kbc, *efq*, are precisely like a carpenter's square. The arms *bk*, *ef*, may be an inch square, and 2 feet 6 inches long; the blades *bc*, *fq*, about $\frac{3}{8}$ of an inch thick, and an inch broad. At *c* and *q* two pieces are fixed at right angles, the distance *cd* being 4 inches. In the face of the arm *bk*, let a groove be ploughed and worked under with a side tool, to a dove-tailed shape, like the section shown at *m*. The under side of the arm *ef* is to be cut to match it like the section shown at *n*. The arm *ef* will now slide along the arm *bk*; in fact, it would be preferable if it were cut like a slide-rule, but carpenters have not tools for effecting this. One inch from *a* (which is opposite to *d*) make a mark, and place a cipher 0. Then from 0 to *k* will be 25 inches; divide this into inches and tenths, and number it from 0 towards *k*. On the arm *ef* at the point opposite to *h* make a mark, and 1 inch from it toward *e* place 25; then divide the space from 25 to *e* into inches and tenths, and number them backward. When this arm is made to slide in the other, and drawn out to measure the length or bung diameter, the number standing opposite the end *k* will denote such length or bung diameter.

To find the content of a large circular vessel, that appears to bulge irregularly, by an odd number of equidistant diameters.

178. Let the vessel be the cask on page 169, and let there be taken 9 diameters, commencing with the head, level with *e*, and proceeding with one between that and D, down to the bottom, which suppose to be 80, 83, 86, 88, 90, 89, 87, 84, and 81 inches, and the depth of the vessel 96 inches, consequently, the common distance of the diameters 12 inches.

Place in a line the letters x^2 $4e^2$ $2o^2$

Under x^2 place the square of the first or top diameter; under $4e^2$ the square of the second diameter; under $2o^2$ the square of the third diameter; under $4e^2$ the square of the fourth diameter; and so on, alternately, to the last, the square of which place under x^2, along with the other extreme. Add together the three columns separately, and multiply that under $4e^2$ by 4; and that under $2o^2$

by 2. Add the three together, multiply by the common interval, and divide by the cone divisor.

EXAMPLE.

x^2	$4e^2$	$2o^2$
6400	6880	7396
6561	7744	8100
	7921	7569
12961	7065	
		23065
	29610	2
	4	
		46130
	118440	
	46130	
	12961	

$177531 \times 12 \div 1059.108 = 2011\frac{1}{2}$ gallons.

This, it will be seen, is merely a modification of the general rule for frustums. For, let the diameters, taken in order, be a, b, c, d, e, &c. Then, taking three at a time, we have $a^2 + 4b^2 + c^2$; $c^2 + 4d^2 + e^2$; $e^2 + 4f^2 + g^2$, &c.; that is, $a^2 + 4b^2 + 2c^2 + 4d^2 + 2e^2 + 4f^2 + g^2$; namely, the square of the extremes, plus 4 times the square of the even diameters, plus twice the square of the remaining odd diameters. By the slide-rule the content may be found by taking it as three successive frustums. The same rule obviously applies to the ullaging of a standing cask.

EXAMPLE.

179. The depth of liquor, in a cask partly filled, is 20 inches; five equidistant diameters, measured from the sur-

face downward, are 28, 27, 26, 24, and 22 inches: required the content.

x^2	$4e^2$	$2o^2$
784	729	676
484	576	2
1268	1305	1352
	4	
	5220	
	1268	
	1352	
	$7840 \times 5 \div 1059.108 = 37$ gallons.	

For ullaging a lying cask, the following rule may be employed.

From 10 times the wet inches, subtract the bung; multiply the remainder by the content, and divide by 8 times the bung; the quotient gives the liquor in the cask; *i. e.*

$$U = \frac{(10\,W - B)\,C}{8\,B}$$

To find the content of vessels whose bases are nearly of an elliptical form, proceed for the area of the base as directed on page 129, and (after multiplying by the common distance of the ordinates,) instead of dividing by 3, multiply by the depth of the vessel, and divide by the *pyramid* divisors, these being equal to 3 times the prism divisors. If the vessel also bulge up the sides, take an odd number of equidistant areas, and proceed as in the last example. And thus may any solid be measured; always observing, that when equidistant *areas* are taken, the *pyramid* divisors must be employed; and when the *squares* of equidistant *diameters* are used, the *cone* divisors must be selected, for the obvious reason that thrice the latter reduce squares to circles.

TIMBER MEASURE.

To find the superficial content of a plank.

Take the length in feet, and the breadth in inches; then divide (by 12) the product of the dimensions. If the board tapers regularly, take half the sum of the end breadths for the mean breadth.

EXAMPLE.

180. A plank 16 feet 6 inches long, is 10 inches broad at one end, and 18 at the other: what is the content?—Here 14 is the mean breadth. Then to 12 of A set 16½, and under 14 is 19¼ square feet.*

---

* It is much to be regretted that the foot is not divided into 100 equal parts instead of 96, as at present. The mode of working duodecimals, though simple enough in itself, often leads to confusion, from the singular names given to the result. Thus, a piece of wood measures 9 feet 5 inches by 3 feet 8 inches, which multiplied together, according to the prescribed rules, gives what are called 34 feet 6 *inches* and 4 *parts*. Now, these inches, as they are termed, are merely twelfths of a *superficial* foot; and these parts, twelfths of such twelfths; that is, duodecimal fractions, each number decreasing in a twelvefold proportion from left to right, as decimal fractions decrease in a tenfold proportion. The name of inches, given to the 6, conveys no kind of idea; for they are neither inches nor feet, but a mixture of the

To find the content of hewn or four-sided timber.

Take the length in feet, the breadth and thickness in inches. Find a mean proportional between the breadth and thickness; then divide (by 12^2) the length multiplied by the square of the mean proportional. If the tree tapers regularly from end to end, find the mean proportional between the mean breadth and thickness.

EXAMPLE.

181. A log of wood is 23 feet 6 inches long, 15 inches thick, and 22 broad: required the content.—Over 15 of D set 15, and under 22 is 18.17, a mean; then over 12 of D set 23½, and over 18.17 is 53$\frac{5}{6}$ solid feet.

182. The length of a piece of timber is 23.8 feet; the breadth at the greater end is 20.18 inches, at the less 16.42 inches; the thickness at the greater end is 14.12 inches, at the less 10.48 inches: required the content.—Here $20.18 + 16.42 = 36.6$, the half of which $= 18.3$ the

---

two, the content being, in reality, 34 square feet, together with another piece, 1 foot long and 6 inches broad; and another, 1 foot long, and $\frac{4}{12}$ of an inch broad; a square foot, in fact, being the *integer*, and the others successive twelfths. For the carpenter the present nomenclature answers well enough, as he perfectly understands that it is a trifle more than 34 feet and a half, which is sufficient for his purpose. But a misconception of the *principle* of duodecimals, carried out by unskilful people, has led to the wildest confusion; for even in some works on arithmetic, designed for the *instruction* of the young, occurs the unimaginable problem of multiplying half-a-crown by half-a-crown; the result of which notable achievement is stated to be 6*s.* 3*d.*

mean breadth; also, $14.12 + 10.48 = 24.6$, the half of which $= 12.3$. Over 12.3 of D set 12.3, and under 18.3 is 15, a mean proportional: then, over 12 of D set 23.8, and over 15 is 37.2.

To find the content of round timber. Take the length in feet, and the girt in inches: then divide (by 12^2) the length multiplied by the square of the quarter girt.

EXAMPLE.

183. Required the content of a tree 48 feet long, the girts at the ends being 60 and 18 inches.—Here 39 is the mean girt, $\frac{1}{4}$ of which $= 9.75$. Then, over 12 of D set 48, and over 9.75 is 31.7 feet nearly.

The above rule gives only about $\frac{11}{14}$ of the true content, but is adopted in practice, as it compensates the purchaser for the waste of timber occasioned by squaring it. The following rule gives the true content very nearly. Divide (by 12^2) twice the length, multiplied by the square of $\frac{1}{5}$ of the girt.

EXAMPLE.

184. Required the content of the last-mentioned tree.—Here $\frac{1}{5}$ of $39 = 7.8$. Hence, over 12 of D set 96 (twice the length) and over 7.8 is 40.56 cubic feet.

But neither this, nor the rule for squared timber, is quite correct, if the tree tapers, but is sufficiently so for all practical purposes.

LAND SURVEYING.

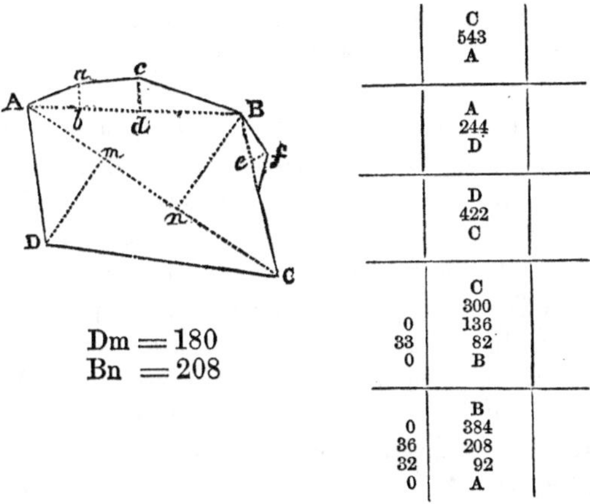

	C	
	543	
	A	
	A	
	244	
	D	
	D	
	422	
	C	
	C	
	300	
0	136	
33	82	
0	B	
	B	
0	384	
36	208	
32	92	
0	A	

Set up poles at A, B, C, and D, so that, standing at A, you can see B and D, the end of the sides whose meeting forms the angle: and so of the others. And suppose the hedge to run on straight, or nearly so, from A to *a*, then to bend and run on straight to *c*, and so on. Form a field-book, as on the right of the diagram, by ruling two lines down the middle of a page; and, in using this, begin at the bottom and write upward, placing the main lines in the middle, and the offsets right or left, as they are on the right or left of the line measured. Then, suppose you commence surveying at A; let your attendant lead the chain toward B; and when he gets it extended, see that he is in a straight line between yourself and B, directing him by a wave of the hand, right or left, according as you wish him to move to one side, or the other. His position being correct, he is to place an arrow in the ground, and walk on till the chain is again extended, when he places another arrow, while you take up the first; and so proceed. But when you arrive at *b*, opposite to *a*, measure *ab*, (at right angles to AB,) with an offset staff, which may be a thin piece of deal 10 links long. Now, suppose from A to *b*, 92 links, and *ab* 32 links; place the 92 in the middle column over A, and 32 on the left hand of it; and so proceed till you arrive at the end of B, which suppose 384; set this down in the field-book, and over it place the letter B, and above this draw a line across the page. The A being placed at the bottom, and B at the top, shows that the intervening numbers are the measures of the AB line. Proceed with the rest in like manner, making the circuit of the field, and returning to A. Then from A, measure the diagonal AC. With these dimensions plot the field from a scale of equal parts, (feather-edged plotting scales are best for this purpose,) and drop the perpendiculars D*m*, B*n*, and from the scale ascertain their lengths. These are set down underneath the diagram, as they are not supposed to have been measured in the field; but if a cross staff, or theodolite, be employed, they are to be taken while proceeding along the diagonal, and set down in the field-book, like offsets, and then the sides AD, DC will not require measuring, supposing there are no offsets on them.—To find the area. Add together D*m*, B*n*, and multiply their sum by AC. For the offsets; for the triangle A *ab*, multiply A*b* by *ba*. For the succeeding trapezoïd, add together *ab*, *cd*, and multiply by *bd*; and so proceed. Then, as, in all these cases, this gives double the area, add the whole together, halve the sum, and divide by the number of links in an acre, viz. 100,000, (that is, point off the 5 right-hand figures,) and the content is in acres and decimal parts; the latter of which being multiplied by 4 and 40, (which need not be set down,) gives the roods and poles. In taking the distances from the field-book, the numbers up the middle column, 92, 208, &c., have to be subtracted from each succeeding, and the offsets, 0, 32, 36, &c., to be added together in pairs. The work will stand thus:—

180	92	116	176	136	210684
208	32	68	36	33	2944
					7888
388	194	928	1056	408	6336
543	276	696	528	408	4488
1164	2944	7888	6336	4488	2)232340
1552					
1940					1.16170
					.64680
210684					25.8720

Content, 1 acre 25 poles.

TRIGONOMETRY AND NAVIGATION.

THE trigonometrical slide is a slide containing the logarithmic sines and tangents, the former of which work to the line D, and the latter to the line A, which lines, as before explained, are also logarithmic. But it is to be recollected, that it is the distances only that are logarithmic, not the numbers; hence, when the slide is laid *evenly* in, then the numbers on A are the natural tangents, and the numbers on D the natural sines of the degrees marked on the slide: when in any other position they are *proportional* to the natural sines and tangents of those degrees; and, therefore, if we set the first term of a proportion over or under the second, then the third will stand over or under the fourth, the fifth over or under the sixth, and so on. In making use of the tangent line, three points may be taken as radius, either the beginning, middle, or end of the slide; but the middle point, marked 45°, will in practice be most convenient. In using the sine slide, two points may be taken as radius, either the beginning or the end of the slide, as may be found necessary for preventing the numbers from overrunning. Having already given several questions under the sector, and Navigation being only an application of Trigonometry, it will be sufficient here to show the mode of working an example or two with the slide-rule, which the student will find infinitely superior for the purpose. Take the tower at page 68. By the sine line, sin. $42\frac{1}{2}$: 200 : : sin. $47\frac{1}{2}$: height,

$$\text{or } \frac{\text{sin.}\, 42\frac{1}{2}}{200} = \frac{\text{sin.}\, 47\frac{1}{2}}{\text{height}}; \text{ therefore—}$$

Under $42\frac{1}{2}$ of the sines bring 200; then under $47\frac{1}{2}$ is 218.2, the height.

By the tangent line, making AC radius—

Rad. or tan. 45° : 200 : : tan. $47\frac{1}{2}$: height;

$$\text{or } \frac{200}{\text{tan. } 45} = \frac{\text{height}}{\text{tan. } 47\frac{1}{2}}.$$

Over 45 of the tangents set 200; then over $47\frac{1}{2}$ is 218.2, the height, (as before.)

NAVIGATION.

North.	South.	Points.	Course.	Complement.	Points.	South.	North.
		$\frac{1}{4}$	2°49′	87°11′	$7\frac{3}{4}$		
		$\frac{1}{2}$	5 37½	84 22½	$7\frac{1}{2}$		
		$\frac{3}{4}$	8 26	81 34	$7\frac{1}{4}$		
N.b.E. N.b.W.	S.b.E. S.b.W.	1	11 15	78 45	7	E.b.S. W.b.S.	E.b.N. W.b.N.
		$1\frac{1}{4}$	14 4	75 56	$6\frac{3}{4}$		
		$1\frac{1}{2}$	16 52½	73 7½	$6\frac{1}{2}$		
		$1\frac{3}{4}$	19 41	70 19	$6\frac{1}{4}$		
NN.E. NN.W.	SS.E. SS.W.	2	22 30	67 30	6	E.S.E. W.S.W.	E.N.E. W.N.W.
		$2\frac{1}{4}$	25 19	64 41	$5\frac{3}{4}$		
		$2\frac{1}{2}$	28 7½	61 52½	$5\frac{1}{2}$		
		$2\frac{3}{4}$	30 56	59 4	$5\frac{1}{4}$		
N.E.b.N. N.W.b.N.	S.E.b.S. S.W.b.S.	3	33 45	56 15	5	S.E.b.E. S.W.b.W.	N.E.b.E. N.W.b.W.
		3	36 34	53 26	$4\frac{3}{4}$		
		$3\frac{1}{2}$	39 22½	50 37½	$4\frac{1}{2}$		
		$3\frac{3}{4}$	42 11	47 49	$4\frac{1}{4}$		
N.E. N.W.	S.E. S.W.	4	45 0	45 0	4	S.E. S.W.	N.E. N.W.
North.	South.	Points.	Complement.	Course.	Points.	South.	North.

PLANE SAILING.

Plane sailing supposes the earth to be a plane, the meridians parallel to each other, and the lengths of degrees everywhere equal; and involves the consideration of four quantities, difference of latitude, nautical distance, departure, and course. Let K (diagram p. 48, disregarding the circle,) denote a point on the earth's surface, and KC its meridian. Draw a line from K to A, and suppose a ship to sail along it from K till she arrive at A; then KA will be the distance sailed; DA, the departure from the meridian; KD, the difference of latitude; and the angle DKA, contained between the meridian and the rhumb sailed on, the course. The difference of latitude is thus represented by a vertical line, the departure by a horizontal one, the distance by the hypothenusal line forming with the other two a right-angled triangle, and the course by the angle included between the difference of latitude and the distance. Then, if we make distance radius, the departure becomes the sine, and the difference of latitude the cosine, of the course; or, if diff. lat. be made radius, departure becomes the tangent of the course.

EXAMPLES.

1. A ship sails 38° S., 255 miles W.: required the diff. of lat. and departure. Complement of 38° = 52°, then—

Sin. 90 : 255 : : sin. 52° : diff. lat. : : sin. 38° : dep.

Under 90 of the sines set 255; then—

Under 52° is 201 miles, diff. lat.
and under 38° is 157 miles, departure.

2. A ship sails from lat. 44° 50′ N. between S. and E. till she has made 64 miles of easting, and is then found to be in lat. 42° 56′ N. : required the course and distance.

$$\begin{array}{r} 44^\circ\ 50' \\ 42\ \ 56 \\ \hline 1\ \ 54 \end{array} = 114 \text{ miles, diff. lat.}$$

As 114 : rad. : : 64 : tan. of course.

Under 114 set 45° of the tangents, then under 64 is 29° 20′, the course.

Again, sin. 29° 20′ : 64 : : sin. 90° : dist.

Under 29° 20′ set 64, then under 90° is 130.6 miles, distance.

3. A ship in lat. 45° 25′ N. sails N.E.b.N. ½ E. till she comes to 46° 55′ N.: required the distance and departure.

N.E.b.N. ½ E. = 39° 22½′, comp. of which = 50° 37½′.

$$\begin{array}{r} 46^\circ\ 55' \\ 45\ \ 25 \\ \hline 1\ \ 30 \end{array} = 90 \text{ miles, diff. lat.}$$

Sin. 50° 37½′ : 90 miles : : sin. 39° 22½′ : 73.8 miles departure : : sin. 90° : 116.4, distance.*

* As the learner is supposed by this time to be familiar with the mode of operation, it will be sufficient for the future to indicate the proportion, without repeating the directions for setting the slide. Thus, in the above instance, under 50° 37½′ set 90 miles, then under 39° 22½′ will be 73.8, and under 90° will be 116.4 miles; and so of all others. When the word rad. occurs as the first or second term, before adjusting the slide run the eye along the proportion to see if the word sin. or tan. follows,

EXAMPLES FOR PRACTICE.

4. A ship sails from lat. 56° 50′ N. on a rhumb between S. and S. W. 126 miles, and is then found to be in lat. 55° 40′: required the course she sailed, and her departure from the meridian.—Ans. Course, 56° 15′; departure, 104.8 miles.

5. A ship in lat. 44° 50′ N. sails S. 29° 20′ E. 130.8 miles: required diff. lat. and departure.—Ans. 64 dep.: 114 diff. lat.

6. A ship in lat. 45° 25′ N. sails N.E.b.N. ½ E. 116.4 miles: required dep., diff. lat., and latitude come to.—Ans. 74 dep.; 90 miles, or 1° 30′ diff. lat; and 46° 55′ N. lat. come to.

7. A ship at sea sails from lat. 34° 24′ N. between N. and W. 124 miles, and is found to have made 86 miles of westing: required the course steered, and diff. of lat., or northing made good.—Ans. Course, 43° 54′; diff. lat. 1° 29′; 35° 53′ N. lat. come to.

8. A ship in lat. 24° 30′ S. sails S.E.b.S. till she has made 96 miles of easting: required the distance sailed, and diff. of lat. made good.—Ans. Diff. lat. 143.7; distance, 172.8; lat. come to, 26° 54′ S.

---

and use the sine or tangent line accordingly. And in every case it will be advisable for the beginner to construct a diagram, as nothing tends so much to make the operation perfectly understood; and what is thoroughly understood at the commencement is seldom afterwards forgotten.

TRAVERSE SAILING.

When a ship sails upon several courses, the zigzag line she describes is called a traverse; and the reducing the courses into one, and thereby finding the course and distance made good upon the whole, is called the resolving of the traverse. For this purpose, construct a table of six columns, in the first of which is the course, and in the second the distance; then find the diff. lat. and dep. for each course, and enter it N. or S., E. or W., as it may be. Add up the columns separately; the difference of the third and fourth will give the diff. of lat., and the diff. of the fifth and sixth, the departure. Then, having obtained the total diff. lat. and dep. which the ship has made, find the corresponding course and distance.

EXAMPLE.

9. A ship from the equator sails N. 48, W. 37, N.W. 18, N.E. 70, N.N.E. 24, and E. 32 miles: required her course, distance, and latitude reached.

Course.	Dist.	Diff. Lat.		Dep.	
		N.	S.	E.	W.
N.	48	48	—	—	—
W.	37	—	—	—	37
N.W.	18	12.72	—	—	13.72
N.E.	70	49.5	—	49.5	—
NN.E.	24	22.18	—	9.18	—
E.	32	—	—	32.	
		132.4		90.68	49.72
				49.72	
				40.96	

First she sails due N., and so will have no departure: therefore place 48 under N.

Her second course is due W., and so she will have no diff. of lat.; therefore place 37 under W.

Her third course is 45°, and therefore her departure and diff. of lat. will be equal. Over 18 place 90° of the sines, and under 45° is 12.72, which place under N. and W.

Her fourth course is also 45°, and therefore her dep. and diff. of lat. will be equal. Over 70 place 90°, and under 45° is 49.5, which place under N. and E.

Her fifth course is 22½, cos. of which = sin. 67½. Over 24 place 90°, then under 22½ is 9.18, her departure, which place under E.; and under 67½ is 22.18, her diff. lat., which place under N.

Her last course is 32 due E., and so she will have no diff. of lat.: therefore place 32 under E.

Add up the three columns. As there is no number standing under S. the diff. of lat. is 132.4 = 2° 12′ N.

Subtract the W. from the E., and the remainder is 40.96 E. for the total departure: then—

As 132.4 : rad. : : 40.96 : tan. 17° 12′, the course.

Again, sin. 17° 12′ : 40.96 : : sin. 90° : 138.6 miles, the distance.

EXAMPLES FOR PRACTICE.

10. A ship from Cape Clear, lat. 51° 25′ N., sails SS.E. ¼ E. 16, E.S.E. 23, S.W.b.W.½ W. 36, W.¾ N. 12, and S.E.b.E. ¼ E. 41 miles: required the equivalent course and distance, and the latitude of the place which the ship has arrived at.—Ans. Course, 18° 12′; distance 62.75 miles; lat. in, 50° 25′ N.

11. From Cape St. Vincent, in lat. 37° 2′ N., a ship sailed S.W.b.S. 49, S.b.E. 56, S.E.b.E. 38, S.W. 84, NN.W. 72, and E.N.E. 24 miles: required the course, distance, and latitude come to.—Ans. Course, 26° 15′; distance, 112; lat. in, 35° 22′ N.

PARALLEL SAILING.

Since the meridians meet at the poles, it follows that the length of a degree on any parallel of latitude diminishes as it recedes from the equator. To ascertain this diminution, when a vessel sails on a parallel of latitude, or changes her longitude only, is the object of parallel sailing. Let FA (diagram, page 48,) represent the earth's semi-axis; FCB, a quadrant of a meridian; B, a point on the equator; C, a point on the meridian, and consequently the arc CB, or angle CAB, the latitude of C; and let the quadrant revolve on AF; then the circles described by the points C, B, or similar parts of them, will be proportional to their radii EC, AB.

Now AB, or AC : EC, or AD : : rad. : cos. CAB;

that is, difference of longitude, or distance between any two meridians on the equator, or parallel described by B : the distance between those meridians on the parallel described by C : : radius : the cosine of the latitude; or the lengths of degrees on different parallels vary as the cosines of the latitudes. Hence, if in any right-angled triangle ADC, the acute angle at the base CAD, be made equal to the latitude, and the length of the base AD equal to the departure, or meridian distance, or distance between any two meridians on a parallel of that latitude;

then the hypothenuse AC will be equal to the arc of the equator, or the difference of longitude corresponding to that meridian distance.

EXAMPLES.

12. Required the number of miles contained in a degree of longitude, in lat. 55 N.

$$\cos. 55^\circ = \sin. 35^\circ.$$

$$\sin. 90^\circ : 60 \text{ miles} :: \sin. 35^\circ : 34.4 \text{ miles}.$$

13. A ship from lat. 42° 52′ N. in long. 9° 17′ W. sails due W. 342 miles: required the longitude come to.

$$\cos. 42^\circ 52' = \sin. 47^\circ 8'.$$

$$\sin. 47^\circ 8' : 342 :: \sin. 90^\circ : 467, \text{ diff. long.}$$

$$\begin{array}{rl} 467 = & 7^\circ 47' \\ & 9\ \ 17 \\ \hline & 17\ \ \ 4 \text{ W. long. come to.} \end{array}$$

14. A ship sailed 224 miles upon a due W. course, and by observation found she had differed her longitude 6° 18′, or 378 miles: required latitude.

$$378 : \sin. 90^\circ :: 224 : \sin. 36^\circ 20';$$

and $\sin. 36^\circ 20' = \cos. 53^\circ 40'$, the latitude required.

15. Two ships in lat. 46° 30′ N., distant asunder 654 miles, sail both directly N. 256 miles: required their distance.

$$\begin{array}{rl} 256 = & 4^\circ 16' \\ & 46\ \ 30 \\ \hline & 50\ \ 46 \text{ N., lat. reached.} \end{array}$$

Then cos. 46° 30′ : 654 miles : : cos. 50° 46′; or sin. 43° 30′ : 654 : : sin. 39° 14′ : 601 miles, the distance.

16. Two ships in lat. 45° 44′ N., distant 846 miles, sail directly N. till the distance between them is 624 miles: required the lat. reached and dist. sailed.

Cos. 45° 44′ = sin. 44° 16′; then 846 : sin. 44° 16′ : : 624 : sin. 31°; and sin. 31° = cos. 39°, lat. come to.

Then 59° 0′
45 44
13 16 = 796 miles, dist. sailed.

EXAMPLES FOR PRACTICE.

17. A ship in lat. 54° 20′ N. sails directly W. on that parallel till she has differed her longitude 12° 45′: required the distance sailed.—Ans. 446 miles.

18. A ship from Cape Finisterre, lat. 42° 52′ N., long. 9° 17′ W., sailed due W. 342 miles: required the longitude come to.—Ans. 17° 4′ W.

19. A ship sails on a certain parallel directly W. 624 miles, and has then differed her longitude 18° 46′, or 1126 miles: required the latitude of the parallel sailed on.—Ans. 56° 20′.

20. A ship from a port in lat. 54° N. sailed due E. 200 miles; then, having run due S. an unknown number of miles, sailed W. 250 miles, and, by observation, found she had arrived at the meridian of the port she sailed from: required the lat. come to, and distance run in the S. direction.—Ans. 42° 43′ lat. come to; 677 miles run.

MIDDLE LATITUDE SAILING.

Middle Latitude Sailing is a composition of plane and parallel sailing, and is used for reducing the departure to miles of longitude. Now, when two places lie not on the same parallel, their difference of longitude, reduced to miles of easting, or westing, if reckoned on the higher parallel, would be too small, and if on the lower parallel, too great. The common way of reducing it is, by taking it, as the name implies, on a parallel midway between the two; which, though not strictly correct, is sufficiently so for most nautical purposes. For the solution of questions of this kind, we have only to place together the two triangles treated of under Plane and Parallel Sailing, and resolve them separately, observing to begin with that in which two parts are given, and then the unknown parts of the other triangle will be easily obtained. See triangle ACK, (diagram, page 48.) By plane sailing, the angle at K is the course; KD, the difference of latitude; DA, the departure; and KA, the distance sailed. By parallel sailing AD is still the departure, or meridian distance, on the parallel midway between the latitude left and latitude reached; CAD, the angle of the middle latitude; and AC, the difference of longitude. The following examples will illustrate the modes of solution.

EXAMPLES.

21. Required the course and distance from the east point of St. Michael's, in lat. 37° 49′ N., long. 25° 11′ W., to Start Point, in lat. 50° 13′ N., long. 3° 38′ W.

50° 13′ N.		25° 11′ W.
37 49 N.		3 38 W.
½)12 24 = 744, diff. lat.		21 33 = 1293, diff. long.
6 12		
37 49		
44 1 mid. lat., complement of which = 45° 59′.		

Then sin. 90° : 1293 : : sin. 45° 59′ : 930, the departure.
744 : rad. : : 930 : tan. 51° 20′, the course = N. 51° 20′ E.
sin. 51° 20′ : 930 : : sin. 90° : 1191 miles, the distance.

22. A ship from Brest, in lat. 48° 23′ N., long. 4° 30′ W., sailed S.W. ¾ W. 238 miles: required the lat. and long. come to.

S.W. ¾ W. = 53° 26′, comp. of which = 36° 34′;

Then

sin. 90° : 238 : : sin. 36° 34′ : 141.8 diff. lat. = 2° 22′.

48° 23′ N.
2 22

46 1 lat. come to.
1 11 = ½ diff. lat.

47 12 mid. lat., comp. of which = 42° 48′.

Then sin. 42° 48′ : 238 : : sin. 53° 26′ : 282 diff. long.

= 4° 42′ W.
4 30 W.

9 12 long. come to.

23. A ship from lat. 17° N., long. 24° 25′ W., sailed N.W. $\frac{3}{4}$ N. till, by observation, her lat. is found to be 28° 34′ N.: required the distance sailed and long. come to.

N.W. $\frac{3}{4}$ N. = 36° 34′, comp. of which = 53° 26′.

28° 34′ N.
17 0 N.

$\frac{1}{2}$)11 34 = 694 diff. lat.

5 47
17 0

22 47 mid. lat., comp. of which = 67° 13′.

Then
sin. 53° 26′ : 694 : : sin. 90° : 864 miles, the distance;

and sin. 67° 13′ : 864 : : sin. 36° 34′ : 558, diff. long.
= 9° 18′ W.
24 25 W.

33 43 long. come to.

EXAMPLES FOR PRACTICE.

24. A ship from lat. 26° 30′ N., long. 45° 30′ W., sailed N.E. $\frac{1}{2}$ N. till her departure was 216 miles: required the distance run, and lat. and long. come to.—Ans. Dist. 341 miles; lat. come to, 30° 53′ N.; long. 41° 24′ W.

25. From lat. 43° 24′ N., long. 65° 39′ W., a ship sailed 246 miles, on a direct course between S. and E., and was then, by observation, in lat. 40° 48′ N.: required

the course, and long. in.—Ans. Course, 50° 40′; long. come to, 61° 23′ W.

26. A ship from Cape St. Vincent, lat. 37° 2′ N., long. 9° 2′ W., sails between S. and W.; the lat. come to is 18° 16′ N., and departure 838 miles: required the course, distance run, and long. come to.—Ans. Course, 36° 40′; dist. 1403 miles; long. come to, 24° 48′ W.

27. A ship from Bordeaux, in lat. 44° 50′ N., 0° 35′ W., sails between the N. and W. 374 miles, and makes 210 miles of easting: required the course, and lat. and long. come to.—Ans. Course, 34° 10′; lat. come to, 49° 59′ N., long. 5° 45′ W.

28. A ship from lat. 54° 56′ N., long. 1° 10′ W., sailed between N. and E. till, by observation, she was found to be in long. 5° 26′ E., and had made 220 miles of easting: required the lat. come to, and course and distance run.—Ans. Lat. come to, 57° 34′ N.; course, 54° 20′; distance, 271 miles.

29. A ship from a port in N. lat. sailed S. E. ¼ S. 438 miles, and differed her long. 7° 28′: required the lat. sailed from and come to.—Ans. Lat. sailed from, 51° 40′; come to, 46° 16′.

TO DETERMINE THE DIFFERENCE OF LONGITUDE MADE GOOD UPON COMPOUND COURSES, BY MIDDLE LATITUDE SAILING.

WITH the several courses and distances find the latitude and departure made good and the ship's present latitude, as in Traverse Sailing. Take the middle latitude between the latitude left and latitude arrived at; then with the departure made by the traverse table, and the middle latitude, find the difference of longitude by Middle Latitude Sailing. In high latitudes this method will be somewhat incorrect, and therefore it will be advisable to employ the more tedious mode of computing the difference of longitude for every separate course, which is most readily done as follows:—Complete the traverse table, as before, to which annex five columns: in the first put the several latitudes the ship is in at the end of each course and distance; in the second, the sums of each consecutive pair of latitudes; and in the third, half the sums, or middle latitude; then find the difference of longitude answering to each separate middle latitude, and its corresponding departure, and place it in the fourth or fifth (namely the east or west) difference of longitude columns, according as the departure is east or west: then the difference of the sums of the east and west columns will be the difference of longitude made good, of the same name as the greater.

EXAMPLE.

30. A ship from lat. 66° 14′ N., long. 3° 12′ E., sails NN. E. ½ E. 46, N. E. ½ E. 28, N. ¾ W. 52, N. E. b. E. ¼ E. 57, and E. S. E. 24 miles : required her course, and longitude in.

Course.	Dist.	Diff. Lat. N.	Diff. Lat. S.	Dep. E. W.	Dep. W.
NN. E. ½ E.	46	40.6	—	21.7	—
N. E. ½ E.	28	17.8	—	21.6	—
N. ¾ W.	52	51.4	—	—	7.6
N. E. b. E. ¼ E.	57	29.3	—	48.9	—
E. S. E.	24	—	9.2	22.2	—
		139.1	9.2	114.4	7.6
		9.2		7.6	
		129.9 = 2° 10′ N.		106.8 E.	

Lat. left. … 66° 14′ N.
Diff. lat. …. 2 10 N.

68 24 lat. in.
1 5 = ½ diff. lat.

67 19 mid. lat. = com. 22° 41′.

130 : rad. : : 106.8 : tan. 39° 25′, the course.
Sin. 39° 25′ : 106.8 : : sin. 90° : 168, distance.
Sin. 22° 41′ : 168 : : sin. 39° 25′ : 277, diff. long.

Long. left…. 3° 12′ E.
Diff. ……… 4 37 E.

7 49 long. come to.

Successive Latitudes.	Sums.	Middle Latitudes.	Diff. Long. E.	Diff. Long. W.
66° 14′				
66 55	133° 9′	66° 34′	54.6	—
67 13	134 8	67 4	55.4	—
68 4	135 17	67 38	—	20
68 33	136 37	68 18	132.3	—
68 24	136 57	68 28	60.5	—
			302.8	20
			20.	
			282.8 = 4° 43′ E.	

Long. left…. 3° 12′ E.
Diff…………. 4 43 E.

7 55 E. long. come to.

(See Example 45.)

EXAMPLES FOR PRACTICE.

31. If a ship sail from the Naze, in lat. 57° 58′ N., long. 7° 3′ E., W. N. W. 24, N. W. ½ W. 16, SS. W. 31, S. ¼ E. 12, and S. W. ¾ S. 20 miles: required her lat. and long.—Ans. Course S. 56° 24′ W.; lat. 57° 20′ N.; long. 5° 15′ E.

32. If a ship sail from the Cape of Good Hope, lat. 34° 29′ S., 18° 23′ E., N. W. 25, N. ½ W. 21, NN. E. ¼ E. 35, N. W. ¾ W. 40, and N. b. E. 18 miles: required her lat. and long.—Ans. Lat. 32° 37′ S., and long. 17° 43′ E.

MERCATOR'S SAILING.

In Mercator's Sailing, so called from the name of its inventor, Gerard Mercator, the earth is conceived to be projected on a plane. In this projection, the meridians are parallel to each other, and, consequently, all places upon it are distorted, and the more so as they approach the poles; but, to compensate for this distortion, the degrees of latitude are everywhere increased in the same proportion as those of longitude; and, consequently, the bearings between places, and the proportions between the latitude, longitude, and nautical distance, will be the same as those on the globe. To examine into this proportion, let us refer again to diagram, page 48. It was shown, in parallel sailing, that any arc described by C is to a similar arc described by B as AD to AC. But AD : AC : : AB : AG; consequently, any arc described by C is to any similar arc described by B as AB is to AG; that is, as radius is to the secant of the latitude. If, therefore, as in Mercator's Projection, the meridians are everywhere equidistant, and, consequently, each parallel of latitude equal to the equator; then the length of any arc, as of a minute, or a degree, on any parallel, is elongated beyond its just proportion, in the same ratio as the secant of the latitude of that parallel exceeds radius. Therefore, to keep up the proportion of northing and southing with that of easting and westing, the length of a minute upon the meridian at any parallel must be increased beyond its just proportion in the ratio of the secant to radius. Consequently, the meridional parts of any given latitude are found by adding together the natural secants of successive minute portions of that latitude; and the smaller these are taken, the more correct will be the table so formed. One sufficient for the purposes of the Slide-Rule is here given.

TABLE OF MERIDIONAL PARTS TO EVERY DEGREE OF THE QUADRANT.

°	MP.	°	MP.	°	MP.	°	MP.	°	MP.	°	MP.	°	MP.	°	PM.	°	MP.
0	0	10	603	20	1225	30	1888	40	2623	50	3474	60	4527	70	5966	80	8375
1	60	11	664	21	1289	31	1958	41	2702	51	3569	61	4649	71	6146	81	8739
2	120	12	725	22	1354	32	2028	42	2782	52	3665	62	4775	72	6335	82	9145
3	180	13	787	23	1419	33	2100	43	2863	53	3764	63	4905	73	6534	83	9606
4	240	14	848	24	1484	34	2171	44	2946	54	3865	64	5039	74	6746	84	10137
5	300	15	910	25	1550	35	2244	45	3030	55	3968	65	5179	75	6970	85	10765
6	361	16	973	26	1616	36	2318	46	3116	56	4074	66	5324	76	7210	86	11532
7	421	17	1035	27	1684	37	2393	47	3203	57	4183	67	5474	77	7467	87	12522
8	482	18	1098	28	1751	38	2468	48	3292	58	4294	68	5631	78	7745	88	13916
9	542	19	1161	29	1819	39	2545	49	3382	59	4409	69	5795	79	8046	89	16300

To return to the diagram. Let the angle DAC be the course; AD the difference of latitude; AC the distance; and DC the departure; then AB being the elongated, or meridional, difference of latitude, AG will be the elongated distance, and BG the elongated departure, that is the real difference of longitude.

Now, AD : DC : : AB : BG; that is, Diff. Lat. : Dep. : : Merid. Diff. Lat. : Diff Long.

And AB : BG : : Rad. : tan. CAD; that is, Merid. Diff. Lat. : Diff. Long. : : Rad. : tan. course.

To find the meridional parts answering to any number of degrees and minutes, take proportional parts of the differences found by subtraction.

EXAMPLES.

33. Required the meridional parts answering to 37° 43′, viz. $37\frac{43}{60}$.

```
38° = 2468
37° = 2393
      ----
        75          Then 2393
        43             +   54
      ----               ----
       225             = 2447 merid. parts for 37° 43′.
      300                ----
      ----
6,0 ) 322,5
      ----
        54 nearly.
      ----
```

34. Required the meridional parts for 27° 58′.

```
28° = 1751
27° = 1684          Then 1751
      ----             -    2
        67               ----
         2             = 1749 merid. parts for 27° 58′.
      ----               ----
6,0 ) 13,4
      ----
         2
      ----
```

35. Required the number of degrees answering to 3625 meridional parts.

3665 = 52°	3625
3569 = 51°	3569
96 = 1° or 60′	56 = ?′

96 : 60′ : : 56 : 35′.

∴ 3625 = 51° 35′.

EXAMPLES IN MERCATOR'S SAILING.

36. Required the course and distance from the east point of the Azores, lat. 37° 49′ N., long. 25′ 11′ W., to Start Point in lat. 50° 13′ N. long. 3° 38′ W.

25° 11′ W.	50° 13′ N. Meridional parts	= 3495
3 38 W.	37 49 N. "	= 2454
21 33 = 1293 miles diff. long.	12 24 = 744 miles diff. lat.	1041 merid. diff. lat.

Then 1041 : rad. : : 1293 : tan. 51° 10′ course, whose comp. = 38° 50′;

and sin. 38° 50′ : 744 : : sin. 90° : 1186, the distance.

Compare this with Example 21.

37. A ship sails from lat. 38° 47′ N., long. 75° 4′ W., 267 miles N. E. b. N.: required the ship's present place.

N. E. b. N. = 33° 45′ course : comp. of which = 56°15′

sin. 91° : 267 : : sin. 56° 15′ : 222 diff. lat. = 3° 42′ N.

38 47 N.

42 29 lat. come to

42° 29′ merid. parts		= 2821
38° 47′	"	= 2528
		293 merid. diff. lat.

Rad. : 293 :: tan. 33° 45′ : 196 diff. long. = 3° 16′ E.

75° 4′ W.
3 16 E.

71 48 long. in.

38. A ship from Nova Scotia, in lat. 45° 20′ N., long. 60° 55′ W., sailed S. E. ¼ S., and, by observation, was found to be in lat. 41° 14′ N.: required the distance sailed, and long. come to.

S. E. ¼ S. = 42° 11′ course, whose comp. = 47° 49′.

45° 20′ merid. parts = 3058
41 14 do. = 2720

4 6 = diff. lat. 338 merid. diff. lat.

Rad. : 338 : : tan. 42° 11′ : 306 diff. long. = 5° 6′.

60° 55′ W.
5 6 E.

55 49 W. long. in.

Sin. 47° 49′ : 246 : : sin. 90° : 332, distance.

EXAMPLES FOR PRACTICE.

39. Required the direct course and distance between the Lizard in lat. 50° 0′ N., and Port Royal, in Jamaica, in lat. 17° 40′ N., differing in long. 70° 46′, Port Royal lying so far to the W. of the Lizard.—Ans. Course, 60° 33′; distance, 3645 miles.

40. Suppose a ship from the Lizard, in lat. 50° N., sails S. 35° 40′ W. 156 miles: required lat. come to, and how much she has altered her longitude.—Ans. Lat. come to, 47° 53′ N.; diff. long. 2° 19′.

41. A ship in lat. 54° 20′ N. sails S. 33° 45′ E., until, by observation, she is found to be in lat. 51° 45′ N.: required the distance sailed, and the diff. long.—Ans. Distance, 186.4 miles: diff. long. 2° 52′ E.

42. A ship from lat. 45° 26′ N. sails between N. and E. 195 miles, and then, by observation, is found to be in lat. 48° 6′ N.: required the direct course, and diff. long.—Ans. Course, N. 34° 52′ E., or N. E. b. N. 1° 7′ E.; diff. long. 2° 43′ E.

43. A ship from lat. 48° 50′ N. sails S. 34° 40′ E., till her diff. long. is 2° 44′: required lat. come to, and distance sailed.—Ans. Diff. lat. 2° 41′; distance 196 miles.

44. A ship from 54° 36′ N. sails S. 42° 33′ W., until she has made 116 miles of departure: required the lat. she is in, her direct distance sailed, and how much she has altered her longitude.—Ans. Lat. come to 52° 30′; distance, 171.5 miles; diff. long. 3° 15′.

TO DETERMINE THE DIFFERENCE OF LONGITUDE MADE GOOD UPON COMPOUND COURSES, BY MERCATOR'S SAILING.

With the several courses and distances, find the latitude and departure made good, and the ship's present latitude, as in traverse sailing. Take the meridional difference of latitude between the latitude left and latitude arrived at. Then, with the course made good by the traverse table, and the meridional difference of latitude, find the difference of longitude by Mercator's Sailing. In high latitudes, this method will be somewhat incorrect; and, therefore, it will be advisable to employ the more tedious mode of computing the difference of longitude for every separate course, which is most readily done as follows:—Complete the traverse table as before, to which annex five columns. In the first, put the several latitudes the ship is in at the end of each course; in the second, the meridional parts corresponding to each latitude; and in the third, the difference of each consecutive pair of meridional parts. Then find the difference of longitude answering to each separate course, and its corresponding meridional difference of latitude, and place it in the fourth or fifth (viz. the east or west) difference of longitude columns, according as the course is east or west; then the difference of the sums of the east and west columns will be the difference of longitude made good, of the same name as the greater.

EXAMPLE.

45. A ship from lat. 66° 14′ N., long. 3° 12′ E., sails NN. E. ½ E. 46, N. E. ½ E. 28, N. ¾ W. 52, N. E. b. E. ¼

E. 57, and E. S. E. 24 miles: required her course and longitude in.

Course.	Dist.	Diff. Lat.		Dep.		Succes-sive Lat.	Merid. Parts.	Merid. Diff.L.	Diff. Long.	
		N.	S.	E.	W.				E.	W.
						66° 14′	5358			
NN.E. ½ E.	46	40.6	—	21.7	—	66 55	5461	103	55.1	—
N.E. ½ E.	28	17.8	—	21.6	—	67 13	5507	46	56.1	—
N. ¾ W.	52	51.4	—	—	7.6	68 4	5642	135	—	20
N.E. b. E. ¼ E.	57	29.3	—	48.9	—	68 33	5720	78	130.1	—
E.S.E.	24	—	9.2	22.2	—	68 24	5695	25	60.4	—
		139.1	9.2	114.4	7.6				301.7	20
		9.2		7.6					20.	
		129.9 = 2°10′N.		106.8 E.					281.7 = 4°42′ E.	

Lat. left 66° 14′ N. merid. parts = 5358
Diff. lat. 2 10 N.

Lat. in 68 24 " = 5695

Merid. diff. lat. 337

130 : rad. :: 106.8 : tan. 39° 25′, the course.
rad. : 337 :: tan. 39° 25′ : 227, diff. long.

Long. left 3° 12′ E.
diff. 4 37 E

7 49 long. come to.

Long. left 3° 12′ E.
4 42 E.

7 54 long. come to.

Compare this with Example 30.

NOTE.—When the course is N. or S., no diff. long. will be made; and when it is due E. or W., find the diff. long. by parallel sailing.

EXAMPLES FOR PRACTICE.

46. A ship from lat. 57° 30′ N., long. 1° 47′ W., sailed SS. E. 48, S. W. b. S. 54, E. b. S. 71, N. E. 63, and W. N. W. 50 miles: required the lat. and long. of the place come to.—Ans. By 1st rule, lat. come to, 56° 50′ N., long. 2′ W.

47. Four days ago, we took our departure from Farohead, in lat. 58° 40′ N., and long. 4° 50′ W., and since have sailed as follows :—N.W. 32, W. 69, W. N.W. 93, W. b. S. 77, S.W. 58, and W. ¾ S. 49 miles : required our present lat. and long.—Ans. By Rule 2, lat. come to, 58° 35′; long. 15° 54′ W.

OBLIQUE SAILING.

Oblique Sailing is the application of oblique-angled plane triangles to the solution of problems at sea; and is particularly useful in going along shore, and surveying coasts and harbours.

EXAMPLES.

48. Coasting along the shore, I saw a cape bear from me NN. E.; then I stood away N.W. b. W. 20 miles, and observed the same cape to bear from me N. E. b. E. : required the distance of the ship from the cape at her last station. See figure, page 72.

Sin. 33° 45′ : 20 :: sin. 78° 45′ : 35.3 miles.

49. A point of land was observed, by a ship at sea, to bear E. b. S.; and after sailing N. E. 12 miles, it was found to bear S. E. b. E. It is required to determine the place of that headland, and the ship's distance from it at the last observation. See figure, page 78.

Sin. 22° 30′ : 12 :: sin. 56° 15′ : 26.1.

50. At noon, Dungeness bore N. b. W., distance 5 leagues; and having run N. W. b. W. 7 knots an hour, at 5 P. M. we were up with Beachy Head : required the distance of Beachy Head from Dungeness.—Ans. 26.6 miles.

WINDWARD SAILING.

Windward Sailing is the method of gaining an intended port by the shortest and most direct method possible, when the wind is in a direction unfavourable to the course the ship ought to steer for that port. In order to attain this point, it is evident that the ship must sail on different tacks; and, therefore, the object of this sailing is, to find the proper courses to be steered on each board, that the vessel may arrive at the intended port with the least delay possible. By the term board is to be understood the shifting of the direction, or alteration of the course. Thus, if a vessel sails on two boards, she shapes out the letter V; if on three boards, the letter N; and so on.

EXAMPLES.

51. A ship is bound to a port 48 miles directly to the windward, the wind being SS.W., which it is intended to reach on two boards; and the ship can lie within 6 points of the wind; required the course and distance on each tack.

Describe a circle, and from the centre, which call A, draw a line in a SS.W. direction, to represent the direction of the wind, and call the lower extremity of this line B, and let it represent the port intended to be reached. Then the wind blowing from B to A, and A being the position of the ship; from A, to the left of the line BA, draw a line, making with it an angle of 6 points, or 67° 30′; this will, of course, be due W. From the centre of the circle A, to the right of the line AB, draw another line, making with AB an angle, like the other, of 67° 30′. This line will be south-east. From the point B, parallel to this last line, draw a line, cutting the one running west, in a point, which call C. Then AC will be the course of the ship on the first board, and CB that on the second. Now, the angles at A and B will be each $67\frac{1}{2}°$, and at C 45°, opposite which is the line AB, 48 miles.

Then, sin. 45° : 48 miles : : sin. $67\frac{1}{2}°$: 62.7 miles, the distance to be sailed on each board; so that she will have to sail 125.4 miles to make 48.

52. The wind at N. W., a ship bound to a port 64 miles to the windward, proposes to reach it on three boards, two on the starboard, and one on the larboard tack, and each within 5 points of the wind: required the course and distance on each tack.

Describe a circle, and from its centre, which call A, draw a line in a N. W. direction, to represent the direction of the wind, and let its upper extremity denote the port intended to be reached, which call B. From A draw two lines, one to the left and the other to the right of the line BA, each making with it an angle of 5 points; consequently, the first will pass through the W. b. S. rhumb, and the second through the N. b. E. Call the lower extremity of the line passing through the S. b. W. rhumb, C; the upper extremity of the other, D. From B draw a line to the right of the line BA, parallel with CA. Bisect BA, in a point, which call E. Draw a line from E to C, parallel with the line DA, and prolong it upward till it cuts the line running right of B, in a point, which call F. Then, in the triangle EAC, the angles at A and C are each 56° 15′, and the angle at E 67° 30′, and the line EA is 32 miles. Therefore, sin. 56° 15′ : 32 : : sin. 67° 30′ : 36.25 miles = AC, BF, CE, or EF, and twice 36.25 = 72.5. Hence, she must first sail W b. S. 36¼ miles, then N. b. E. 72½ miles, then W. b. S. 36¼ miles.

It may be here observed, that whatever number of boards it may be found expedient a ship should make, the *sum* of the distances on each tack will be the same as if the place had been reached on two boards only.

53. A ship is bound to a port 26 miles directly to windward (the wind being N. E.,) which it is intended to reach on two boards, the first being on the larboard tack, and the ship can lie within 6 points of the wind: required the course and distance on each tack.—Ans. Course on the larboard tack, E. S. E.; on the starboard, NN. W.; distance on each board, 34 miles, nearly.

54. The wind at N. ½ E., a ship is bound to a port bearing NN. E., distance 68 miles, which it is proposed to make at four boards; the coast, which is to westward, trends NN. E. also; so that the ship must go about as soon as she reaches the straight line joining the ports: required the course and distance on each board, the ship making her way good within 6 points of the wind.—Ans. Course on the larboard tack, E. N. E. ½ E.; on the starboard, N. W. b. W. ½ W.; first and third distances, 47.8 miles; second and fourth distances, 37.2 miles.

55. A ship close hauled within 5 points of the wind, and making 1 point of leeway, is bound to a port bearing SS. W., distant 54 miles, the wind being S. b. E.; it is intended to make the port at three boards, the first of which must be on the larboard tack, in order to avoid a reef of rocks: required the course and distance on each

tack.—Ans. Course on the larboard tack, S. W. b. W.; on the starboard, E. b. S.; distances on the larboard tack, each 37.45 miles; distance on the starboard tack, 42.4 miles.

CURRENT SAILING.

When a ship sails exactly with the current, her velocity will, of course, be accelerated; and, when in due opposition to the current, it will be retarded by the difference of the velocities of the wind and stream. When she is urged by the wind in one direction, and by the current in another, her course, agreeably to the law influencing all bodies acted upon simultaneously by two forces, will lie in the diagonal of the parallelogram formed by those forces; that is, will be the third side of a triangle of which the drift of the current and the action of the wind form the other two, the angle between them being known.

N.B. That point of the compass to which a current runs is called its setting, and its rate per hour is called its drift.

EXAMPLES.

56. A ship sails by the compass directly S. 96 miles, in a current that sets E. 45 miles in the same time: required the ship's true course and distance.

Describe a circle, and from its centre, which call A, draw a line in a south direction, and make it equal to 96 from a scale of equal parts, and call the lower extremity B. From the point B, in an easterly direction, draw a line equal to 45, from the same scale, and call its extremity C. Join AC. The angle BAC will be the course, and C the point at which the ship will have arrived. Then,

96 : rad. : : 45 : tan. 25° 7′, the ship's course = SS. E. 2° 6′, easterly.

And, sin. 25° 7′ : 45 : : sin. 90° : 105.9 miles, distance sailed.

57. A ship has made by the reckoning N. ½ W. 20 miles, but, by observation, it is found that, owing to a current, she has actually gone NN. E. 28 miles: required the setting and drift of the current in the time which the ship had been running.—Ans. Setting, N. 64° 48′ E., drift, 14.1 miles.

58. A ship from a port in lat. 42° 52′ N., sailed S. b. W. ½ W. 17 miles in 7 hours, in a current setting between the N. and W.; and then the same port bore E. N. E., and the ship's latitude, by observation, was 42° 42′ N.: required the setting and drift of the current.—Ans. Setting, 71° 55′, drift, 2.9 knots an hour.

59. A ship, bound from Dover to Calais, lying 21 miles to the S. E. b. E. ½ E., and the flood-tide setting N. E. ½ E. 2½ miles an hour: required the course she must steer, and the distance run by the log, at 6 knots an hour, to reach her port.—Ans. Course, 39° 14′. Distance to be run 19.4 miles.

60. From a ship, in a current, steering W. S. W. 6 miles an hour by the log, a rock was seen at 6 in the evening, bearing S. W. ½ S. 20 miles. The ship was lost on the rock at 11 P. M.: required the setting and drift of the current.—Ans. Setting, S. 75° 10′ E., drift 3.11 miles per hour.

OF A SHIP'S JOURNAL.

A journal is a register of transactions occurring on board a ship, and should contain a particular detail of every thing relative to the navigation of the vessel—as the courses, winds, currents, &c.—that her situation may be known at any instant at which it may be required. The computations made to determine the place of a ship from the courses and distances run in 24 hours, are called a day's work; and the latitude and longitude of a ship deduced therefrom, are called the latitude and longitude in, *by account*, or, *by dead reckoning*, in contradistinction to the latitude and longitude as determined by observation. At the time of leaving land, the bearing of some known place is to be observed, and its distance found, either by observation, or by taking its bearing at two different times, from two different places, and determining its distance accordingly. The log-book, which is to contain a daily transcript from the log-board, is to be divided into 7 columns. In the first, put the hours; in the second and third, the knots and fathoms sailed per hour; in the fourth, the courses; in the fifth, the winds; in the sixth, the leeway; and in the seventh, any remark that may be thought necessary. It is better, however, to omit the leeway column, and, on transcribing from the log-board, to make the proper allowance, and to enter the amended courses only, in the log-book. After this, allow for the variation, and bring them into a traverse table. Find the ship's distance, difference of latitude, and departure, by plane sailing; then, by Mercator's, or middle latitude sailing, find the difference of longitude, and enter it accordingly. The following specimen will furnish an idea; but the present work being intended principally to show the instrumental modes of computation, the student is referred to works exclusively on Navigation, for more complete information upon the subject.

Hours.	Knots.	Fathoms.	Courses.	Winds.	REMARKS. *Monday, Aug.* 2, 1847.
2	6	...	W.b.N.	N.E.	A fresh steady breeze. P.M.
4	6				
6	6	...	...	...	Spoke H. M. Ship *Alligator*.
8	6				
10	6				
12	6	...	...	...	Fair Weather.
2	6	...	...	...	A light on the larboard bow. A.M.
4	6				
6	6				
8	7	...	...	...	Fresh breezes and clear.
10	8	...	...	...	Lizard N.E. ½ E., distant 5 leagues, lat. 49° 57′ N., long. 5° 15′ W.
12	8	...	...	...	Do. Weather. Variation 2½ points W.

Course.	Distance.	Diff. Lat.	Departure.	N. Latitude by Account.	N. Latitude by Observation.	Diff. Long.	W. Longitude by Account.	W. Longitude by Observation.
S. 48° 40′ W.	28	18	21	49° 39′		33′ W.	5° 48′	

The departure is taken from the Lizard, at 10 A. M. The bearing is N. E. ½ E., distance 15 miles. Now, the opposite point is S.W. ½ W., and the variation, 2½ points, being allowed to the left hand, because it is westerly, gives SS.W., the true bearing of the ship from the Lizard; so it will be SS.W., 15 miles. The course the ship has been going, is W. b. N., which, corrected for variation, is W. S.W. ½ W.; and the distance run from 10 A. M. to noon, is 16 miles. Now, insert these in a traverse table, as under, and find the diff. lat. and departure to each course and distance by plane sailing. Hence the diff. lat. and departure made good will be obtained, with which the course and distance from the Lizard will be determined. Then, with the departure and middle latitude find the difference of longitude.

TRAVERSE TABLE.

Courses.	Distance.	Diff. Lat.		Departure.	
		N.	S.	E.	W.
SS.W.	15		13.9		5.7
W.S.W. ½ W.	16		4.6		15.3
S. 48° 40′ W.	28		18.5		21.

Lat. left 49° 57′ N.
Diff. lat. 18 S.

49 39 lat. in.
9 = ½ diff. lat.

49 48 mid. lat. = comp. 40° 12′.

18.5 : rad. : : 21 : tan. 48° 40′, the course.
sin. 48° 40′ : 21 : : sin. 90° : 28, the distance.
sin. 40° 12′ : 28 : : sin. 48° 40′ : 32.6 diff. long. = 33′ nearly.

Long. left 5° 15′ W.
Diff. long. 33 W.

5 48 W. long. come to.

Hours.	Knots.	Fathoms.	Courses.	Winds.	REMARKS. *Tuesday, Aug.* 3, 1847.	
2	8	...	W.b.S.	N.E.	A fresh gale.	P.M.
4	8	4				
6	8	6	...	...	A sail upon the lee-beam.	
8	9	...	W.S.W.			
10	9					
12	9					
2	9	...	...	...		A.M.
4	9					
6	9					
8	9	...	S.W.b.W.	...	Do. Weather, and clear.	
10	9	6				
12	9	4	...	...	Variation $2\frac{1}{2}$ points W.	

Course.	Distance.	Diff. Lat.	Departure.	N. Latitude by		Diff. Long.	W. Longitude by	
				Account.	Observation.		Account.	Observation.
S. 38° 52′ W.	212	165	133	46° 54′	46° 56′	200	9° 8′	

The several courses being corrected for variation, the diff. lat. and departure, answering to each course and distance, will be as under.

Courses.	Distance.	Diff. Lat.		Departure.	
		N.	S.	E.	W.
S.W. ½ W.	50		31.7		38.6
S.W. ½ S.	108		83.5		68.5
SS.W. ½ W.	56		49.4		26.4
S. 38° 52′ W.	212		164.6		133.5

Yesterday's lat. 49° 39′ N.
Diff. lat............ 2 45 S.

46 54 N. lat. in.
1 22 = ½ diff. lat.

48 16 mid. lat. = comp. 41° 44′.

165 : rad. : : 133 : tan. 38° 52′, the course.

Sin. 38° 52′ : 133 : : sin. 90° : 212, the distance.

Sin. 41° 44′ : 212 : : sin. 38° 52′ : 200 diff. long. = 3° 20′.

Yesterday's long. 5° 48′ W.
Diff. long........... 3 20 W.

9 8 W. long. in by account.

RECAPITULATION.

QUESTIONS ON TABLE I.

1. THE diameter of a circle is 9 inches: what is the circumference?—Ans. 28.27.

2. What is the side of an equal square?—Ans. 7.97.

3. The circumference of a circle is 23 inches: what is the diameter?—Ans. 7.32.

4. What is the side of an inscribed square?—Ans. 5.17.

5. The side of a square is 18: what is the diameter of an equal circle?—Ans. 20.3.

6. What is the circumference of an equal circle?—Ans. 63.8.

7. The area of a circle is 24: what is the area of its inscribed square?—Ans. 15.27.

8. The area of a square is 24: what is the area of its inscribed circle?—Ans. 18.85.

QUESTIONS ON TABLE II.

9. The diameter of a circle is 25 inches: what is the side of an inscribed equilateral triangle?—Ans. 21.65.

10. Of an inscribed pentagon?—Ans. 14.69.

11. Of a circumscribed decagon?—Ans. 8.12.

12. Of an inscribed undecagon?—Ans. 7.04.

13. Of a circumscribed dodecagon?—Ans. 6.69.

14. The diameter of a circle is 12 inches: what is the side of a square inscribed in it?—Ans. 8.48.

15. A circle, whose diameter is 9¼ inches, has a regular hexagon surrounding it: what is the length of each side? —Ans. 5.33.

16. An octagonal tower measures 7 feet along each side: what will be the diameter of a circle surrounding it?—Ans. 18.3.

17. What is the length of the longest line that can be drawn within a dodecagon, each of whose sides is 7 feet? —Ans. 26.13.

QUESTIONS ON TABLE III.

18. The side of an equilateral triangle is 7: what is the area?—Ans. 21.2.

19. The side of a regular pentagon is 7: what is the area?—Ans. 84.3.

20. The side of a regular heptagon is 7: what is the area?—Ans. 178.

21. The side of a regular nonagon is 7: what is the area?—Ans. 302.9.

22. The side of a regular dodecagon is 6: what is the area?—Ans. 403.

23. The side of a regular hexagon is 47 inches: how many square feet does it contain?—Ans. 39.86.

24. What is the area, in square yards, of an undecagon whose side measures 17.9 feet?—Ans. 334.

QUESTIONS ON TABLE IV.

25. A bullet, let fall from a balloon, was half a minute before it struck the earth: how high was the aëronaut at the moment it was dropped?—Ans. 4825 yds., or 2¾ miles.

26. When the same balloon had attained an altitude of 4 miles, or 7040 yards, another bullet was let fall: how many seconds was it in descending?—Ans. 36¼ seconds.

27. What is the height of a precipice, if a stone is 7 seconds in falling from the top to the bottom?—Ans. 788 feet.

28. A string, with a bullet at the end, being suspended from a hook in the ceiling, is found to vibrate 64 times per minute: what is the distance from the hook to the centre of the bullet?—Ans. 34.4 inches.

29. How often will a pendulum, 100 inches long, vibrate per minute?—Ans. 37½ times.

30. A revolving pendulum shapes out 52 cones in a minute: determine its length?—Ans. 13 inches.

31. The diameter of a circle is 60 inches: what is the area?—Ans. 2827½ square inches.

32. The diameter of a sphere is 9 inches: what is the convex surface?—Ans. 25.45 square feet.

33. The circumference of a sphere is 12 inches: what is the surface?—Ans. 45.8 square inches.

34. What is the diagonal of a square whose side measures 15.3 inches?—Ans. 21.63.

35. A cube measures 9 inches along the side: what will be the diagonal of the face, and what of the cube?—Ans. 12.72 diagonal of the face; 15.58 diagonal of the cube.

QUESTIONS ON TABLE V.

36. The diameter of a circle is 9 inches: how many square inches does it contain?—Ans. 63.6.

37. The diameter of a circle is 44 inches: how many square feet does it contain?—Ans. 10.55.

38. The side of a square is 17.5 inches: required the area in square feet.—Ans. 2.126.

39. The diameters of an ellipse are 12 and 10 feet: required the area in square yards.—Ans. 10.47.

40. What is the area in square yards of a cycloid, whose generating circle has a diameter of 3 feet?—Ans. 2.356.

41. Required the surface of a cylinder, in square feet, the circumference of which is 29 inches, and height 42 inches.—Ans. 8.46.

42. The diameter of a sphere is 73 feet: what is the surface in square rods?—Ans. 61.5 nearly.

QUESTIONS ON TABLE VI.

43. A vessel in the shape of a square prism is 40 inches deep, and 12 inches square: how many solid feet does it contain?—Ans. 3.33, or $3\frac{1}{3}$ feet.

44. An inverted octagonal pyramid measures 5 inches along each side at the top, and is 13 inches deep: how many gallons will it contain?—Ans. 1.88 gallons.

45. A dodecagonal pyramid measures 6 inches along each side at the bottom, and is 16.6 inches high: how many solid feet does it contain?—Ans. 1.29 feet.

46. A cone of ice is 50 inches in perpendicular height; the diameter of its base is 10 inches: required its weight.—Ans. 43.9 lbs.

47. A hollow sphere, 11 inches in diameter, is filled with tallow: required its weight.—Ans. 23.1 lbs.

48. How much gunpowder would fill the same?—Ans. 23.4 lbs.

49. The axes of an oblate spheroid are 20 and 22 inches: how many gallons will it hold?—Ans. 18.2 gallons.

50. The axes of a prolate spheroid are 20 and 22 inches: how many gallons will it hold?—Ans. 16.6 gallons.

51. The axes of an elliptic cylinder of solid gold are 4

and 9 inches; its depth 3 inches: what is its weight?—Ans. 60 lbs. avoir.

52. The axes of an elliptic cone of brass are 4 and 9 inches; its depth 8 inches: required its weight.—Ans. 22.85 lbs.

53. A paraboloid of zinc is 14 inches high; the radius of its base 5 inches: required its weight.—Ans. 146 lbs. nearly.

54. A parabolic spindle of silver is 23 inches long; its diameter 8 inches: what is its weight?—Ans. 233.6 lbs. avoir.

55. The length of a cask is 45 inches, the bung diameter 36, and the head diameter 30 inches: required the content for each of the four varieties.

Ans.	1st variety		148.37	gallons.
	2d	"	147.76	"
	3d	"	139.96	"
	4th	"	139.2	"

QUESTIONS ON TABLE VII.

56. A sphere of platinum weighs 32 lbs.: required its diameter.—Ans. 4.28 inches.

57. A solid globe of gold weighs 40 lbs.: what is its diameter?—Ans. 4.78 inches.

58. A sphere 5 inches diameter is filled with quicksilver: required its weight.—Ans. 32.4 lbs.

59. Required the diameter of a pound rocket.—Ans. 1.67 in.

60. The internal diameter of rockets is usually $\frac{2}{3}$ of their external: what then is the internal diameter of a 6 lb. rocket?—Ans. 2 inches.

61. A globe of ice weighs 10 lbs.: required its diameter.—Ans. 8.3 inches.

QUESTIONS ON TABLE VIII.

62. A sphere contains 100 cubic inches: required its diameter.—Ans. 5.75 inches.

63. The solidity of a sphere is 180: what is its diameter?—Ans. 7.

64. What is its circumference?—Ans. 22.

65. The solidity of an octahedron is 9: what is the length of each of its sides?—Ans. 2.68.

66. The moon is distant 240 thousand miles from the earth, and the time of her complete revolution is about 27½ days: at what distance would she go round in a week?—Ans. 96 thousand miles.

67. The distance of Venus is 68 millions of miles: how many weeks does she consume in traversing her orbit?—Ans. 32 weeks.

68. Vesta performs her revolution in about 47½ lunar months: required her distance.—Ans. 225 million miles.

69. Juno is distant 253 millions of miles; how many days are consumed in her revolution?—Ans. 1590.

70. At what distance would a planet require to be placed to revolve round the sun in 2 years?—Ans. 151 millions of miles.

71. Ceres is distant 263, Pallas 265 millions of miles from the sun: how many years is each employed in her circuit?—Ans. Ceres, 4.6 years; Pallas, 4.67 years.

---

72. I have 3 balls, weighing 1 lb., 2 lbs., and 7½ lbs. respectively; the smallest is 3 inches diameter: required the diameter of the other two.—Ans. 3.78, and 5.87 inches.

73. A cone weighing 74 lbs. is 24.6 inches high, and

10 inches diameter at the base: required the size of a similar cone, weighing 100 lbs.—Ans. 27.19 inches high; 11.05 diam.

74. I have two similarly shaped casks, the dimensions of one are, length 54, head 34.8, bung 44.8, and middle diameter 83 inches; the other holds $2\frac{2}{5}$ times as much: required the dimensions.—Ans. L. 72.3; H. 46.6; B. 59.98; and M. 111.1.

75. Out of a sheet of metal, of uniform thickness, a piece is cut in the shape of a regular decagon, each of whose sides measures 7 inches; its weight is found to be $8\frac{1}{4}$ lbs.; a similar piece is cut from the same sheet, and weighs $23\frac{3}{5}$ lbs.: what is the length of each of its sides? —Ans. 11.83 inches.

76. Find, by the rule, the cube root of 141.—Ans. 5.2.

77. The frustum of a nonagonal pyramid measures 3 inches along each side at top, and 4 at bottom, and the depth is 10 inches; into this I put a sphere of brass weighing $\frac{4}{5}$ of what the water required to fill the vessel would weigh: what is the diameter of the sphere?—Ans. 4.1 inches.

78. A cast-iron cannon-ball weighs 38 lbs.,: required its circumference.—Ans. 20.4 inches.

79. A tap 2 inches in diameter will empty a cask in $53\frac{1}{2}$ minutes: what must be the size of one to empty it in an hour and $53\frac{1}{2}$ minutes?—Ans. 1.373 inches.

80. A has a globe of lead 4 inches diameter; B, a globe of copper of the same weight: what is its diameter? —Ans. 4.34 inches.

81. The tinker, mentioned at page 162, succeeded in making a similar vessel to contain 20 gallons: required its dimensions.—Ans. Depth, 13.51 inches; bottom diameter, 16.96; top, 28.27.

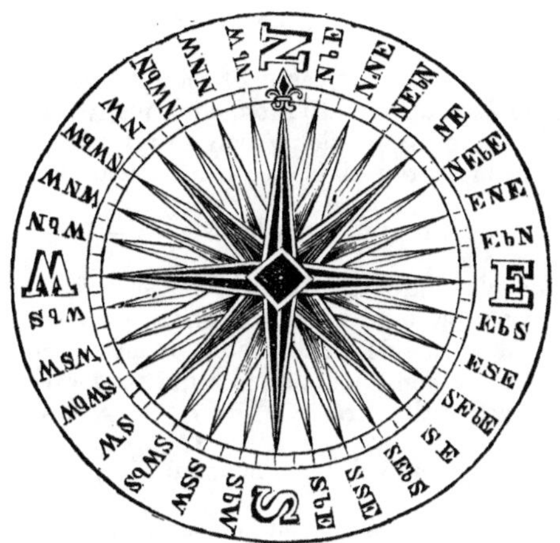

APPENDIX.

During the sale of the first few hundred copies of the present impression, it has been found that the omission of the Compass has proved a great inconvenience; it is, therefore, now supplied, as above.

Most of the operations of the Slide-Rule have been exhibited at pp. 91, 92, &c. The principal of them may be more concisely shown as follows:—Let a and A denote any two logarithmic distances taken on the A line; b and B any two equal distances on the B line; and so on. Then a varies as b as c as d^2, and A as B as C as D^2; e as d^3, and E as D^3; a^3 as e^2 as d^6, and A^3 as E^2 as D^6.

From these an immense variety of combinations may be formed; some of them more curious than useful. The former class the student can investigate for himself; of the latter kind the following are of constant occurrence.

(1.) $a : b :: A : B$, whence $B = \dfrac{bA}{a}$.

Of this class are all cases of simple proportion, including multiplication, division, and many formulæ for surfaces. In multiplication a being unity, in division A; and, for surfaces, a a divisor, b length, and A breadth.

$$(2.)\qquad a : b :: d^2 : c, \text{ whence } c = \frac{bd^2}{a}.$$

Of this class are the formulæ for surfaces and solids, when divisors are used instead of gauge points. For surfaces d will be a side, or diameter, or mean proportional between two dimensions, and b a quantity varying with the boundary of the surface.

$$(3.)\qquad a : b :: d^2 : c, \text{ whence } b = \frac{ca}{d^2}.$$

Of this class are the formulæ for surfaces, in which d is a gauge point, c length, and a breadth.

$$(4.)\qquad c : d^2 :: C : D^2, \text{ whence } C = \frac{cD^2}{d^2}.$$

Of this class are the formulæ for accelerated motion, and for exhibiting the relations of similar plane figures to each other; for finding the areas of surfaces, and the content of solids; d being a gauge point: c, in surfaces, a variable quantity—in solids, length, height, or depth; and D a diameter, side, or mean proportional between two dimensions.

$$(5.)\qquad e : d^3 :: E : D^3, \text{ whence } E = \frac{eD^3}{d^3}.$$

Of this class are the formulæ for determining the dimensions of spheres from their weight, or solidity; and for exhibiting the relations of similar solids to each other.

$$(6.)\qquad a^3 : e^2 :: A^3 : E^2, \text{ whence } E = e\sqrt{\frac{A^3}{a^3}}.$$

The formula for determining the distance of a planet from its periodical revolution, and conversely.

The principle of rules containing inverted lines is shown as follows:—Let a, b, and c, denote any logarithmic distance on the A, B, and C lines; and, in lieu of D, let an inverted line A be laid down, so that unity upon it coincides with the extremity of the C line. Then the value of the same distance upon this line will be $\frac{1}{A}$; but if it be drawn aside until some other number r fall under the extremity, then its value will be $\frac{r}{A}$; and $\therefore$ we shall have $\frac{r}{A} : c :: a : b$, whence $b = \frac{Aca}{r}$, where r is a constant divisor, and A, c, a, any three numbers.

Solutions of the more difficult Questions.

Example 161, page 160.—1 : 6 :: 3 : 18, the depth of the entire cone; hence $\frac{2}{3}$ of the depth is cut off; $\therefore \left(\frac{2}{3}\right)^3$ or $\frac{8}{27}$ of the solidity is cut off, and the remaining frustum is $\frac{19}{27}$; $\frac{4}{5}$ of this is $\frac{76}{135}$, which added to $\frac{8}{27} = \frac{116}{135}$;

$$\text{hence } 135 : 18^3 :: 116 : ?^3$$

135 E : 18 D :: 116 E : 17.12 D, the distance from the surface of the water to the bottom of the cone: hence 17.12 — 12 = 5.12, the depth of the water.

Ex. 162.—2 : 18 :: 3 : 27, the height of the entire pyramid; hence $\frac{1}{3}$ of the height is cut off; $\therefore \left(\frac{1}{3}\right)^3$ or $\frac{1}{27}$ of the whole is cut off, and the remaining frustum is $\frac{26}{27}$; $\frac{1}{3}$ of this is $\frac{26}{81}$; so each person will have $\frac{26}{81}$, and $\frac{3}{81}$ will be for wast● The various bulks will therefore be as 3, 29, 55, and 81.

Hence $81 : 27^3 :: 55 : ?^3 :: 29 : ?^3 :: 3 : 9^3$;

$81 E : 27 D :: 55 E : 23.73 D :: 29 E : 19.17 D :: 3 E : 9 D$; then $27 - 23.73 = 3.27$; $23.73 - 19.17 = 4.56$; $19.17 - 9 = 10.17$.

Ex. 172.—4 lb. : 3^3 :: 108 lb. : $?^3$ $4 E : 3 D :: 108 E : 9$ inches D, the diameter of the globe; to find the content of which in gallons, the globe gauge point is 23; divide, then, by 23^2, 9 times 9^2.

$23 D : 9 C :: 9 D : 1.37$ gallons C;

$12 + 1.37 = 13.37$ gallons, the quantity virtually put into the vessel.

Again, 5 : 20 :: 15 : 60, the height of the entire pyramid; to find the content of which, in gallons, the pentagonal pyramid gauge point is 21.98; divide, then, by 21.98^2, 60 times 15^2.

$21.98 D : 60 C :: 15 D : 27.92$ gallons C;

$27.92 - 13.37 = 14.55$, the content of the pyramidal segment above the surface of the water; then $27.92 : 60^3 :: 14.55 : ?^3$

$27.92 E : 60 D :: 14.55 E : 48.28$ inches D, the height of the segmental pyramid above the water; $\therefore 48.28 - 40 = 8.28$, the depth of the vessel unoccupied.

Ex. 173.— As $\frac{1}{5}$ of the diameter is to be left, $\left(\frac{1}{5}\right)^3$ or $\frac{1}{125}$ of the solidity will be left, and $\frac{124}{125}$ will be turned down; $\therefore$ each will turn down $\frac{31}{125}$; the various bulks will therefore be as 1, 32, 63, 94, and 125.

Hence $125 : 10^3 :: 94 : ?^3 :: 63 : ?^3 :: 32 : ?^3 :: 1 : 2^3$;

$125 E : 10 D :: 94 E : 9.09 D :: 63 E : 7.96 D :: 32 E : 6.35 D :: 1 E : 2 D$.

Ex. 174.—To find, in pints, the contents of a globe whose diameter is 3.6 inches. The pint gauge point for globes is 8.13; divide, then, by 8.13^2, 3.6 times 3.6^2.

$$8.13\,D : 3.6\,C :: 3.6\,D : .703\,C;$$

$$.703 + \frac{7}{9} = .703 + .777 = 1.48;$$

then $.703 : 3.6^3 :: 1.48 : ?^3$ $.703\,E : 3.6\,D :: 1.48\,E : 4.615\,D$, the diameter; and $113\,A : 355\,B :: 4.615\,A : 14.49\,B$, the circumference.

Ex. 175.—Let the diameters be 30 and 50. The round or conic gauge point for gallons is 46; the content, therefore, by formula 17, page 137, is

$$\frac{12}{46^2}\left\{\begin{array}{l}30^2 = ?\\ 50^2 = ?\\ 80^2 = ?\end{array}\right. \quad \text{viz.} \quad \frac{C\,12}{D\,46}\left\{\begin{array}{l}30 = 5.1\\ 50 = 14.2\\ 80 = 36.2\end{array}\right.$$

$$\overline{55.5} \text{ gallons,}$$

the content of a vessel whose depth is 12 inches, and diameters 30 and 50. Then, since the depth remains unaltered, the content will vary as the squares of the diameters;

$$\text{hence } 55.5 \text{ gals.} : \left\{\begin{array}{c}30^2\\ 50^2\end{array}\right\} :: 14 \text{ gals.} : ?^2$$

$$55.5\,C : 30\,D :: 14\,C : 15.06\,D, \text{ bottom diameter;}$$

$$55.5\,C : 50\,D :: 14\,C : 25.1\,D, \text{ top diameter.}$$

Round Timber.

Instead of using the quarter girt, as mentioned at page 176, it will be preferable to take the *whole* girt, and *four* times the divisor; that is, putting L length in feet, g girt in inches, then the content by the common method will be $\frac{Lg^2}{48^2}$; thus, in question 183, $48\,D : 48\,C :: 39\,D : 31.7$

feet *C*. To find the true content the formula will be $\frac{Lg^2}{42.53^2}$; thus, in question 184, 42.53 *D* : 48 *C* :: 39 *D* : $40\frac{1}{2}$ cubic feet *C*.

Casks.

The following exhibits the formulæ for the four varieties under the simplest form :—

1st var. Fr. Pro. Sphd. $\frac{L(H^2+2.B^2)}{32.54^2}$ 2d var. Fr. Par. Spin. $\frac{L(H^2+2.B^2-\frac{1}{10}\text{ of }(2\text{diff.})^2}{32.54^2}$

3d var. Fr. Two Parb. $\frac{L(H^2+B^2)}{26.6^2}$ 4th var. Fr. Two Co. $\frac{L(H^2+B^2+\overline{H+B}|^2)}{46^2}$

It will be found a great improvement to the rule to copy the formulæ at pp. 136, 137, 138, on the back of one of the slides.

June, 1848.

THE END.

STEREOTYPED BY L. JOHNSON & CO.
PHILADELPHIA.

PUBLICATIONS

OF

HENRY CAREY BAIRD,

SUCCESSOR TO E. L. CAREY,

S. E. CORNER OF MARKET AND FIFTH STREETS,

PHILADELPHIA.

SCIENTIFIC AND PRACTICAL.

THE PRACTICAL MODEL CALCULATOR,

For the Engineer, Machinist, Manufacturer of Engine Work, Naval Architect, Miner, and Millwright. By Oliver Byrne, Compiler and Editor of the Dictionary of Machines, Mechanics, Engine Work and Engineering, and Author of various Mathematical and Mechanical Works. Illustrated by numerous Engravings. To be published in Twelve Parts, at Twenty-five Cents each, forming, when completod, One large Volume, Octavo, of nearly six hundred pages.

It will contain such calculations as are met with and required in the Mechanical Arts, and establish models or standards to guide practical men. The Tables that are introduced, many of which are new, will greatly economize labour, and render the every-day calculations of the *practical man* comprehensive and easy. From every single calculation given in this work numerous other calculations are readily modelled, so that each may be considered the head of a numerous family of practical results.

The examples selected will be found appropriate, and in all cases taken from the actual practice of the present time. Every rule has been tested by the unerring results of mathematical research, and confirmed by experiment, when such was necessary.

The Practical Model Calculator will be found to fill a vacancy in the library of the practical working-man long considered a requirement. It will be found to excel all other works of a similar nature, from the great extent of its range, the exemplary nature of its well-selected examples, and from the easy, simple, and systematic manner in which the model calculations are established.

NORRIS'S HAND-BOOK FOR LOCOMOTIVE ENGINEERS AND MACHINISTS;

Comprising the Calculations for Constructing Locomotives. Manner of setting Valves, &c. &c. By Septimus Norris, Civil and Mechanical Engineer. In One Volume, 12mo., with illustrations.. $1.50

A COMPLETE TREATISE ON TANNING, CURRYING AND EVERY BRANCH OF LEATHER-DRESSING.

From the French and from original sources. By CAMPBELL MORFIT, one of the Editors of the "Encyclopedia of Chemistry," Author of "Chemistry Applied to the Manufacture of Soap and Candles," and other Scientific Treatises. Illustrated with several hundred Engravings. Complete in One Vol., royal 8vo. (In press.)

This important treatise will be issued from the press at as early a day as the duties of the editor will permit, and it is believed that in no other branch of applied science could more signal service be rendered to American Manufacturers.

The publisher is not aware that in any other work heretofore issued in this country, more space has been devoted to this subject than a single chapter; and in offering this volume to so large and intelligent a class as American Tanners and Leather Dressers, he feels confident of their substantial support and encouragement.

THE PRACTICAL COTTON-SPINNER AND MANUFACTURER; Or, The Manager's and Overseer's Companion.

This works contains a Comprehensive System of Calculations for Mill Gearing and Machinery, from the first moving power through the different processes of Carding, Drawing, Slabbing, Roving, Spinning, and Weaving, adapted to American Machinery, Practice, and Usages. Compendious Tables of Yarns and Reeds are added. Illustrated by large Working-drawings of the most approved American Cotton Machinery. Complete in One Volume, octavo...$3.50

This edition of Scott's Cotton-Spinner, by OLIVER BYRNE, is designed for the American Operative. It will be found intensely practical, and will be of the greatest possible value to the Manager, Overseer, and Workman.

THE PRACTICAL METAL-WORKER'S ASSISTANT,

For Tin-Plate Workers, Brasiers, Coppersmiths, Zinc-Plate Ornamenters and Workers, Wire Workers, Whitesmiths, Blacksmiths, Bell Hangers, Jewellers, Silver and Gold Smiths, Electrotypers, and all other Workers in Alloys and Metals. By CHARLES HOLTZAPPFEL. Edited, with important additions, by OLIVER BYRNE. Complete in One Volume, octavo............$4.00

It will treat of Casting, Founding, and Forging; of Tongs and other Tools; Degrees of Heat and Managemnet of Fires; Welding; of Heading and Swage Tools; of Punches and Anvils; of Hardening and Tempering; of Malleable Iron Castings, Case Hardening, Wrought and Cast Iron. The management and manipulation of Metals and Alloys, Melting and Mixing. The management of Furnaces, Casting and Founding with Metallic Moulds, Joining and Working Sheet Metal. Peculiarities of the different Tools employed. Processes dependent on the ductility of Metals. Wire Drawing, Drawing Metal Tubes, Soldering. The use of the Blowpipe, and every other known Metal-Worker's Tool. To the works of Holtzappfel, OLIVER BYRNE has added all that is useful and peculiar to the American Metal-Worker.

THE MANUFACTURE OF IRON IN ALL ITS VARIOUS BRANCHES:

To which is added an Essay on the Manufacture of Steel, by Frederick Overman, Mining Engineer, with one hundred and fifty Wood Engravings. A new edition. In One Volume, octavo, five hundred pages..$5.00

We have now to announce the appearance of another valuable work on the subject which, in our humble opinion, supplies any deficiency which late improvements and discoveries may have caused, from the lapse of time since the date of "Mushet" and "Schrivenor." It is the production of one of our transatlantic brethren, Mr. Frederick Overman, Mining Engineer; and we do not hesitate to set it down as a work of great importance to all connected with the iron interest; one which, while it is sufficiently technological fully to explain chemical analysis, and the various phenomena of iron under different circumstances, to the satisfaction of the most fastidious, is written in that clear and comprehensive style as to be available to the capacity of the humblest mind, and consequently will be of much advantage to those works where the proprietors may see the desirability of placing it in the hands of their operatives.—*London Morning Journal.*

A TREATISE ON THE AMERICAN STEAM-ENGINE.

Illustrated by numerous Wood Cuts and other Engravings. By Oliver Byrne. In One Volume, royal 8vo. (In press.)

PROPELLERS AND STEAM NAVIGATION:

With Biographical Sketches of Early Inventors. By Robert Macfarlane, C. E., Editor of the "Scientific American." In One Volume, 12mo. Illustrated by over Eighty Wood Engravings...75 cts.

The object of this "History of Propellers and Steam Navigation" is twofold. One is the arrangement and description of many devices which have been invented to propel vessels, in order to prevent many ingenious men from wasting their time, talents, and money on such projects. The immense amount of time, study, and money thrown away on such contrivances is beyond calculation. In this respect, it is hoped that it will be the means of doing some good.—*Preface.*

MATHEMATICS FOR PRACTICAL MEN:

Being a Common-place Book of Principles, Theorems, Rules, and tables, in various Departments of Pure and Mixed Mathematics, with their applications, especially to the pursuits of Surveyors, Architects, Mechanics, and Civil Engineers. With numerous Engravings. By Olinthus Gregory, LL.D., F. R. A. S...$1.50

Only let men awake, and fix their eye, one while on the nature of things, another while on the application of them to the use and service of mankind.—*Lord Bacon.*

PRACTICAL SERIES.

The volumes in this series are published in duodecimo form, and the design is to furnish to Artisans, for a moderate sum, Hand-books of the different Arts and Manufactures, in order that they may be enabled to keep pace with the improvements of the age.

There have already appeared—

THE AMERICAN MILLER AND MILLWRIGHT'S ASSISTANT. $1.
THE TURNER'S COMPANION. 75 cts.
THE PAINTER, GILDER, AND VARNISHER'S COMPANION. 75 cts.
THE DYER AND COLOUR-MAKER'S COMPANION. 75 cts.
THE BUILDER'S COMPANION. $1.
THE CABINET-MAKER'S COMPANION. 75 cts.

The following, among others, are in preparation:—

A TREATISE ON A BOX OF INSTRUMENTS. By THOMAS KENTISH.
THE PAPER-HANGER'S COMPANION. By J. ARROWSMITH.

THE AMERICAN MILLER AND MILLWRIGHT'S ASSISTANT:

By WILLIAM CARTER HUGHES, Editor of "The American Miller," (newspaper,) Buffalo, N. Y. Illustrated by Drawings of the most approved Machinery. In One Volume, 12mo.........$1

The author offers it as a substantial reference, instead of speculative theories, which belong only to those not immediately attached to the business. Special notice is also given of most of the essential improvements which have of late been introduced for the benefit of the Miller.—*Savannah Republican.*

The whole business of making flour is most thoroughly treated by him.—*Bulletin.*

A very comprehensive view of the Millwright's business.—*Southern Literary Messenger.*

THE TURNER'S COMPANION:

Containing Instructions in Concentric, Elliptic, and Eccentric Turning. Also, various Plates of Chucks, Tools, and Instruments, and Directions for using the Eccentric Cutter, Drill, Vertical Cutter, and Circular Rest; with Patterns and Instructions for working them. Illustrated by numerous Engravings. In One Volume, 12mo...75 cts.

The object of the Turner's Companion is to explain in a clear, concise, and intelligible manner, the rudiments of this beautiful art.—*Savannah Republican.*

There is no description of turning or lathe-work that this elegant little treatise does not describe and illustrate.—*Western Lit. Messenger.*

THE PAINTER, GILDER, AND VARNISHER'S COMPANION:

Containing Rules and Regulations for every thing relating to the arts of Painting, Gilding, Varnishing, and Glass Staining; numerous useful and valuable Receipts; Tests for the detection of Adulterations in Oils, Colours, &c., and a Statement of the Diseases and Accidents to which Painters, Gilders, and Varnishers are particularly liable; with the simplest methods of Prevention and Remedy. In one vol. small 12mo., cloth. 75cts.

Rejecting all that appeared foreign to the subject, the compiler has omitted nothing of real practical worth.—*Hunt's Merchant's Magazine.*

An excellent *practical work*, and one which the practical man cannot afford to be without.—*Farmer and Mechanic.*

It contains every thing that is of interest to persons engaged in this trade. —*Bulletin.*

This book will prove valuable to all whose business is in any way connected with painting.—*Scott's Weekly.*

Cannot fail to be useful.—*N. Y. Commercial.*

THE BUILDER'S POCKET COMPANION:

Containing the Elements of Building, Surveying, and Architecture; with Practical Rules and Instructions connected with the subject. By A. C. Smeaton, Civil Engineer, &c. In one volume, 12mo. $1.

Contents:—The Builder, Carpenter, Joiner, Mason, Plasterer, Plumber, Painter, Smith, Practical Geometry, Surveyor, Cohesive Strength of Bodies, Architect.

It gives, in a small space, the most thorough directions to the builder, from the laying of a brick, or the felling of a tree, up to the most elaborate production of ornamental architecture. It is scientific, without being obscure and unintelligible, and every house-carpenter, master, journeyman, or apprentice, should have a copy at hand always.—*Evening Bulletin.*

Complete on the subjects of which it treats. A most useful practical work. —*Balt. American.*

It must be of great practical utility.—*Savannah Republican.*

To whatever branch of the art of building the reader may belong, he will find in this something valuable and calculated to assist his progress.—*Farmer and Mechanic.*

This is a valuable little volume, designed to assist the student in the acquisition of elementary knowledge, and will be found highly advantageous to every young man who has devoted himself to the interesting pursuits of which it treats.—*Va. Herald.*

THE DYER AND COLOUR-MAKER'S COMPANION:

Containing upwards of two hundred Receipts for making Colors, on the most approved principles, for all the various styles and fabrics now in existence; with the Scouring Process, and plain Directions for Preparing, Washing-off, and Finishing the Goods. In one volume, small 12mo., cloth. 75 cts.

This is another of that most excellent class of practical books, which the publisher is giving to the public. Indeed we believe there is not, for manufacturers, a more valuable work, having been prepared for, and expressly adapted to their business.—*Farmer and Mechanic.*

It is a valuable book.—*Otsego Republican.*

We have shown it to some practical men, who all pronounced it the completest thing of the kind they had seen—*N. Y. Nation.*

THE CABINET-MAKER AND UPHOLSTERER'S COMPANION:

Comprising the Rudiments and Principles of Cabinet Making and Upholstery, with familiar instructions, illustrated by Examples, for attaining a proficiency in the Art of Drawing, as applicable to Cabinet Work; the processes of Veneering, Inlaying, and Buhl Work; the art of Dyeing and Staining Wood, Ivory, Bone, Tortoise-shell, etc. Directions for Lackering, Japanning, and Varnishing; to make French Polish; to prepare the best Glues, Cements, and Compositions, and a number of Receipts particularly useful for Workmen generally, with Explanatory and Illustrative Engravings. By J. Stokes. In one volume, 12mo., with illustrations. 75 cts.

THE PAPER-HANGER'S COMPANION:

In which the Practical Operations of the Trade are systematically laid down; with copious Directions Preparatory to Papering; Preventions against the effect of Damp in Walls; the various Cements and Pastes adapted to the several purposes of the Trade; Observations and Directions for the Panelling and Ornamenting of Rooms, &c. &c. By James Arrowsmith. In One Volume, 12mo. (In press.)

THE FRUIT, FLOWER, AND KITCHEN GARDEN.

By Patrick Neill, L.L.D.

Thoroughly revised, and adapted to the climate and seasons of the United States, by a Practical Horticulturist. Illustrated by numerous Engravings. In one volume, 12mo. $1.25.

HOUSEHOLD SURGERY; OR, HINTS ON EMERGENCIES.

By J. F. South, one of the Surgeons of St. Thomas's Hospital. In one volume, 12mo. Illustrated by nearly fifty Engravings. $1.25.

CONTENTS:

The Doctor's Shop.—Poultices, Fomentations, Lotions, Liniments, Ointments, Plasters.

Surgery.—Blood-letting, Blistering, Vaccination, Tooth-drawing, How to put on a Roller, Lancing the Gums, Swollen Veins, Bruises, Wounds, Torn or Cut Achilles Tendon, What is to be done in cases of sudden Bleeding from various causes, Scalds and Burns, Frost-bite, Chilblains, Sprains, Broken Bones, Bent Bones, Dislocations, Ruptures, Piles, Protruding Bowels, Wetting the Bed, Whitlow, Boils, Black-heads, Ingrowing Nails, Bunions, Corns, Sty in the Eye, Blight in the Eye, Tumours in the Eyelids, Inflammation on the Surface of the Eye, Pustules on the Eye, Milk Abscesses, Sore Nipples, Irritable Breast, Breathing, Stifling, Choking, Things in the Eye, On Dress, Exercise and Diet of Children, Bathing, Infections, Observations on Ventilation.

HOUSEHOLD MEDICINE.

In one volume, 12mo. Uniform with, and a companion to, the above. (In immediate preparation.)

THE ENCYCLOPEDIA OF CHEMISTRY, PRACTICAL AND THEORETICAL:

Embracing its application to the Arts, Metallurgy, Mineralogy, Geology, Medicine, and Pharmacy. By JAMES C. BOOTH, Melter and Refiner in the United States Mint; Professor of Applied Chemistry in the Franklin Institute, etc.; assisted by CAMPBELL MORFIT, author of "Chemical Manipulations," etc. Complete in one volume, royal octavo, 978 pages, with numerous wood cuts and other illustrations. $5.

It covers the whole field of Chemistry as applied to Arts and Sciences. * * * As no library is complete without a common dictionary, it is also our opinion that none can be without this Encyclopedia of Chemistry.—*Scientific American.*

A work of time and labour, and a treasury of chemical information.—*North American.*

By far the best manual of the kind which has been presented to the American public.—*Boston Courier.*

An invaluable work for the dissemination of sound practical knowledge.—*Ledger.*

A treasury of chemical information, including all the latest and most important discoveries.—*Baltimore American.*

At the first glance at this massive volume, one is amazed at the amount of reading furnished in its compact double pages, about one thousand in number. A further examination shows that every page is richly stored with information, and that while the labours of the authors have covered a wide field, they have neglected or slighted nothing. Every chemical term, substance, and process is elaborately, but intelligibly, described. The whole science of Chemistry is placed before the reader as fully as is practicable, with a science continually progressing. * * Unlike most American works of this class, the authors have not depended upon any one European work for their materials. They have gathered theirs from works on Chemistry in all languages, and in all parts of Europe and America; their own experience, as practical chemists, being ever ready to settle doubts or reconcile conflicting authorities. The fruit of so much toil is a work that must ever be an honour to American Science.—*Evening Bulletin.*

PERFUMERY; ITS MANUFACTURE AND USE:

With Instructions in every branch of the Art, and Receipts for all the Fashionable Preparations; the whole forming a valuable aid to the Perfumer, Druggist, and Soap Manufacturer. Illustrated by numerous Wood-cuts. From the French of Celnart, and other late authorities. With Additions and Improvements by CAMPBELL MORFIT, one of the Editors of the "Encyclopedia of Chemistry." In one volume, 12mo., cloth. $1.

A TREATISE ON A BOX OF INSTRUMENTS,

And the Slide Rule, with the Theory of Trigonometry and Logarithms, including Practical Geometry, Surveying, Measuring of Timber, Cask and Malt Gauging, Heights and Distances. By Thomas Kentish. In One Volume, 12mo. (In press.)

STEAM FOR THE MILLION.

An Elementary Outline Treatise on the Nature and Management of Steam, and the Principles and Arrangement of the Engine. Adapted for Popular Instruction, for Apprentices, and for the use of the Navigator. With an Appendix containing Notes on Expansive Steam, &c. In One Volume, 8vo...37½ cts.

SYLLABUS OF A COMPLETE COURSE OF LECTURES ON CHEMISTRY:

Including its Application to the Arts, Agriculture, and Mining, prepared for the use of the Gentlemen Cadets at the Hon. E. I. Co.'s Military Seminary, Addiscombe. By Professor E. Solly, Lecturer on Chemistry in the Hon. E. I. Co.'s Military Seminary. Revised by the Author of "Chemical Manipulations." In one volume, octavo, cloth. $1.25.

The present work is designed to occupy a vacant place in the libraries of Chemical text-books. It is admirably adapted to the wants of both TEACHER and PUPIL; and will be found especially convenient to the latter, either as a companion in the class-room, or as a remembrancer in the study. It gives, at a glance, under appropriate headings, a classified view of the whole science, which is at the same time compendious and minutely accurate; and its wide margins afford sufficient blank space for such manuscript notes as the student may wish to add during lectures or recitations.

The almost indispensable advantages of such an impressive aid to memory are evident to every student who has used one in other branches of study. Therefore, as there is now no Chemical Syllabus, we have been induced by the excellencies of this work to recommend its republication in this country; confident that an examination of the contents will produce full conviction of its intrinsic worth and usefulness.—*Editor's Preface.*

ELECTROTYPE MANIPULATION:

Being the Theory and Plain Instructions in the Art of Working in Metals, by Precipitating them from their Solutions, through the agency of Galvanic or Voltaic Electricity. By CHARLES V. WALKER, Hon. Secretary to the London Electrical Society, etc. Illustrated by Wood-cuts. In one volume, 24mo., cloth. From the thirteenth London edition. 62 cts.

PHOTOGENIC MANIPULATION:

Containing the Theory and Plain Instructions in the Art of Photography, or the Productions of Pictures through the Agency of Light; including Calotype, Chrysotype, Cyanotype, Chromatype, Energiatype, Anthotype, Amphitype, Daguerreotype, Thermography, Electrical and Galvanic Impressions. By GEORGE THOMAS FISHER, Jr., Assistant in the Laboratory of the London Institution. Illustrated by wood-cuts. In one volume, 24mo., cloth. 62 cts.

MATHEMATICS FOR PRACTICAL MEN:

Being a Common-Place Book of Principles, Theorems, Rules, and Tables, in various departments of Pure and Mixed Mathematics, with their Applications; especially to the pursuits of Surveyors, Architects, Mechanics, and Civil Engineers, with numerous Engravings. By OLINTHUS GREGORY, L. L. D. $1.50.

Only let men awake, and fix their eyes, one while on the nature of things, another while on the application of them to the use and service of mankind. —*Lord Bacon.*

AN ELEMENTARY COURSE OF INSTRUCTION ON ORDNANCE AND GUNNERY:

Prepared for the use of the Midshipmen at the Naval School. By JAMES H. WARD, U. S. N. In one volume, octavo. $1.50.

SHEEP-HUSBANDRY IN THE SOUTH:

Comprising a Treatise on the Acclimation of Sheep in the Southern States, and an Account of the different Breeds. Also, a Complete Manual of Breeding, Summer and Winter Management, and of the Treatment of Diseases. With Portraits and other Illustrations. By Henry S. Randall. In One Volume, octavo..$1.25

ELWOOD'S GRAIN TABLES:

Showing the value of Bushels and Pounds of different kinds of Grain, calculated in Federal Money, so arranged as to exhibit upon a single page the value at a given price from *ten cents to two dollars* per bushel, of any quantity from *one pound to ten thousand bushels.* By J. L. Elwood. A new Edition. In One Volume, 12mo.. $1

To Millers and Produce Dealers this work is pronounced by all who have it in use, to be superior in arrangement to any work of the kind published—and *unerring accuracy in every calculation may be relied upon in every instance.*

☞ A reward of Twenty-five Dollars is offered for an error of one cent found in the work.

MISS LESLIE'S COMPLETE COOKERY.

Directions for Cookery, in its Various Branches. By Miss Leslie. Forty-second Edition. Thoroughly Revised, with the Addition of New Receipts. In One Volume, 12mo, half bound, or in sheep..$1

In preparing a new and carefully revised edition of this my first work on cookery, I have introduced improvements, corrected errors, and added new receipts, that I trust will on trial be found satisfactory. The success of the book (proved by its immense and increasing circulation) affords conclusive evidence that it has obtained the approbation of a large number of my countrywomen; many of whom have informed me that it has made practical housewives of young ladies who have entered into married life with no other acquirements than a few showy accomplishments. Gentlemen, also, have told me of great improvements in the family table, after presenting their wives with this manual of domestic cookery, and that, after a morning devoted to the fatigues of business, they no longer find themselves subjected to the annoyance of an ill-dressed dinner.—*Preface.*

MISS LESLIE'S TWO HUNDRED RECEIPTS IN FRENCH COOKERY.

A new Edition, in cloth..25 cts.

TABLES OF LOGARITHMS FOR ENGINEERS AND MACHINISTS:

Containing the Logarithms of the Natural Numbers, from 1 to 100000, by the help of Proportional Differences. And Logarithmic Sines, Cosines, Tangents, Co-tangents, Secants, and Co-secants, for every Degree and Minute in the Quadrant. To which are added, Differences for every 100 Seconds. By Oliver Byrne, Civil, Military, and Mechanical Engineer. In One Volume, 8vo. cloth..$1

TWO HUNDRED DESIGNS FOR COTTAGES AND VILLAS, &c. &c.

Original and Selected. By Thomas U. Walter, Architect of Girard College, and John Jay Smith, Librarian of the Philadelphia Library. In Four Parts, quarto..........................$10

ELEMENTARY PRINCIPLES OF CARPENTRY.

By Thomas Tredgold. In One Volume, quarto, with numerous Illustrations..$2.50

A TREATISE ON BREWING AND DISTILLING,

In One Volume, 8vo. (In press.)

FAMILY ENCYCLOPEDIA

Of Useful Knowledge and General Literature; containing about Four Thousand Articles upon Scientific and Popular Subjects. With Plates. By John L. Blake, D. D. In One Volume, 8vo, full bound..$5

SYSTEMATIC ARRANGEMENT OF COKE'S FIRST INSTITUTES OF THE LAWS OF ENGLAND.

By J. H. Thomas. Three Volumes, 8vo, law sheep.........$12

AN ACCOUNT OF SOME OF THE MOST IMPORTANT DISEASES OF WOMEN.

By Robert Gooch, M. D. In One Volume, 8vo, sheep...$1.50

STANDARD ILLUSTRATED POETRY.

THE TALES AND POEMS OF LORD BYRON:

Illustrated by HENRY WARREN. In One Volume, royal 8vo. with 10 Plates, scarlet cloth, gilt edges..............................$5
Morocco extra..$7

It is illustrated by several elegant engravings, from original designs by WARREN, and is a most splendid work for the parlour or study.—*Boston Evening Gazette.*

CHILDE HAROLD; A ROMAUNT BY LORD BYRON:

Illustrated by 12 Splendid Plates, by WARREN and others. In One Volume, royal 8vo., cloth extra, gilt edges..................$5
Morocco extra..$7

Printed in elegant style, with splendid pictures, far superior to any thing of the sort usually found in books of this kind.—*N. Y. Courier.*

THE FEMALE POETS OF AMERICA.

By RUFUS W. GRISWOLD. A new Edition. In One Volume, royal 8vo. Cloth, gilt...$2.50
Cloth extra, gilt edges...$3
Morocco super extra...$4.50

The best production which has yet come from the pen of Dr. GRISWOLD, and the most valuable contribution which he has ever made to the literary celebrity of the country.—*N. Y. Tribune.*

THE LADY OF THE LAKE:

By SIR WALTER SCOTT. Illustrated with 10 Plates, by CORBOULD and MEADOWS. In One Volume, royal 8vo. Bound in cloth extra, gilt edges...$5
Turkey morocco super extra............... ....$7

This is one of the most truly beautiful books which has ever issued from the American press.

LALLA ROOKH; A ROMANCE BY THOMAS MOORE:

Illustrated by 13 Plates, from Designs by CORBOULD, MEADOWS, and STEPHANOFF. In One Volume, royal 8vo. Bound in cloth extra, gilt edges...$5
Turkey morocco super extra.......................................$7

This is published in a style uniform with the "Lady of the Lake."

THE POETICAL WORKS OF THOMAS GRAY:

With Illustrations by C. W. RADCLIFF. Edited with a Memoir, by HENRY REED, Professor of English Literature in the University of Pennsylvania. In One Volume, 8vo. Bound in cloth extra, gilt edges..........$3.50
Turkey morocco super extra..........$5.50

It is many a day since we have seen issued from the press of our country a volume so complete and truly elegant in every respect. The typography is faultless, the illustrations superior, and the binding superb.—*Troy Whig.*

We have not seen a specimen of typographical luxury from the American press which can surpass this volume in choice elegance.—*Boston Courier.*

It is eminently calculated to consecrate among American readers, (if they have not been consecrated already in their hearts,) the pure, the elegant, the refined, and, in many respects, the sublime imaginings of THOMAS GRAY.—*Richmond Whig.*

THE POETICAL WORKS OF HENRY WADSWORTH LONGFELLOW:

Illustrated by 10 Plates, after Designs by D. HUNTINGDON, with a Portrait. Ninth Edition. In One Volume, royal 8vo. Bound in cloth extra, gilt edges..........$5
Morocco super extra..........$7

This is the very luxury of literature—LONGFELLOW'S charming poems presented in a form of unsurpassed beauty.—*Neal's Gazette.*

POETS AND POETRY OF ENGLAND IN THE NINETEENTH CENTURY.

By RUFUS W. GRISWOLD. Illustrated. In One Volume, royal 8vo. Bound in cloth..........$3
Cloth extra, gilt edges..........$3.50
Morocco super extra..........$5

Such is the critical acumen discovered in these selections, that scarcely a page is to be found but is redolent with beauties, and the volume itself may be regarded as a galaxy of literary pearls.—*Democratic Review.*

THE TASK, AND OTHER POEMS.

By WILLIAM COWPER. Illustrated by 10 Steel Engravings. In One Volume, 12mo. Cloth extra, gilt edges..........$2
Morocco ex:ra..........$2

"The illustrations in this edition of Cowper are most exquisitely designed and engraved."

THE FEMALE POETS OF GREAT BRITAIN.

With Copious Selections and Critical Remarks. By FREDERIC ROWTON. With Additions by an American Editor, and finely engraved Illustrations by celebrated Artists. In One Volume, royal 8vo. Bound in cloth extra, gilt edges.....................$5
Turkey morocco..$7

Mr. ROWTON has presented us with admirably selected specimens of nearly one hundred of the most celebrated female poets of Great Britain, from the time of Lady Juliana Bernes, the first of whom there is any record, to the Mitfords, the Hewitts, the Cooks, the Barretts, and others of the present day.—*Hunt's Merchants' Magazine.*

SPECIMENS OF THE BRITISH POETS.

From the time of Chaucer to the end of the Eighteenth Century. By THOMAS CAMPBELL. In One Volume, royal 8vo. (In press.)

THE POETS AND POETRY OF THE ANCIENTS:

By WILLIAM PETER, A. M. Comprising Translations and Specimens of the Poets of Greece and Rome, with an elegant engraved View of the Coliseum at Rome. Bound in cloth......$3
Cloth extra, gilt edges...$3.50
Turkey morocco super extra......................................$5

It is without fear that we say that no such excellent or complete collection has ever been made. It is made with skill, taste, and judgment.—*Charleston Patriot.*

THE POETICAL WORKS OF N. PARKER WILLIS.

Illustrated by 16 Plates, after designs by E. LEUTZE. In One Volume, royal 8vo. A new Edition. Bound in cloth extra, gilt edges..$5
Turkey morocco super extra......................................$7

This is one of the most beautiful works ever published in this country.—*Courier and Inquirer.*

Pure and perfect in sentiment, often in expression, and many a heart has been won from sorrow or roused from apathy by his earlier melodies. The illustrations are by LEUTZE,—a sufficient guarantee for their beauty and grace. As for the typographical execution of the volume, it will bear comparison with any English book, and quite surpasses most issues in America.—*Neal's Gazette.*

The admirers of the poet could not have his gems in a better form for holi day presents.—*W. Continent.*

MISCELLANEOUS.

ADVENTURES OF CAPTAIN SIMON SUGGS;

And other Sketches. By JOHNSON J. HOOPER. With Illustrations. 12mo, paper..50 cts.
Cloth...62 cts.

AUNT PATTY'S SCRAP-BAG.

By Mrs. CAROLINE LEE HENTZ, Author of "Linda." 12mo.
Paper covers..50 cts.
Cloth...62 cts.

BIG BEAR OF ARKANSAS;

And other Western Sketches. Edited by W. T. PORTER. In One Volume, 12mo, paper....................................50 cts.
Cloth...62 cts.

COMIC BLACKSTONE.

By GILBERT ABBOT A' BECKET. Illustrated. Complete in One Volume. Cloth..75 cts.

GHOST STORIES.

Illustrated by Designs by DARLEY. In One Volume, 12mo, paper covers...50 cts.

MODERN CHIVALRY; OR, THE ADVENTURES OF CAPTAIN FARRAGO AND TEAGUE O'REGAN.

By H. H. BRACKENRIDGE. Second Edition since the Author's death. With a Biographical Notice, a Critical Disquisition on the Work, and Explanatory Notes. With Illustrations, from Original Designs by DARLEY. Two volumes, paper covers $1.00
Cloth or sheep..$1.25

COMPLETE WORKS OF LORD BOLINGBROKE:

With a Life, prepared expressly for this Edition, containing Additional Information relative to his Personal and Public Character, selected from the best authorities. In Four Volumes, 8vo. Bound in cloth..$6.00
In sheep..7.50

CHRONICLES OF PINEVILLE.

By the Author of "Major Jones's Courtship." Illustrated by Darley. 12mo, paper..50 cts.
Cloth..62 cts.

GILBERT GURNEY.

By Theodore Hook. With Illustrations. In One Volume, 8vo., paper..50 cts.

MEMOIRS OF THE GENERALS, COMMODORES, AND OTHER COMMANDERS,

Who distinguished themselves in the American Army and Navy, during the War of the Revolution, the War with France, that with Tripoli, and the War of 1812, and who were presented with Medals, by Congress, for their gallant services. By Thomas Wyatt, A. M., Author of "History of the Kings of France." Illustrated with Eighty-two Engravings from the Medals. 8vo.
Cloth gilt..$2.00
Half morocco..$2.50

GEMS OF THE BRITISH POETS.

By S. C. Hall. In One Volume, 12mo., cloth............$1.00
Cloth, gilt..$1.25

VISITS TO REMARKABLE PLACES:

Old Halls, Battle Fields, and Scenes Illustrative of striking passages in English History and Poetry. By William Howitt. In Two Volumes, 8vo, cloth......................................$3.50

NARRATIVE OF THE ARCTIC LAND EXPEDITION.

By CAPTAIN BACK, R. N. In One Volume, 8vo, boards...$1.50

THE MISCELLANEOUS WORKS OF WILLIAM HAZLITT.

Including Table-talk; Opinions of Books, Men, and Things; Lectures on Dramatic Literature of the Age of Elizabeth; Lectures on the English Comic Writers; The Spirit of the Age, or Contemporary Portraits. Five Volumes, 12mo., cloth......$5.00
Half calf..$6.25

FLORAL OFFERING.

A Token of Friendship. Edited by FRANCES S. OSGOOD. Illustrated by 10 beautiful Bouquets of Flowers. In One Volume, 4to, muslin, gilt edges..$3.50
Turkey morocco super extra......................................$5.50

THE HISTORICAL ESSAYS,

Published under the title of "Dix Ans D'Etude Historique," and Narratives of the Merovingian Era; or, Scenes in the Sixth Century. With an Autobiographical Preface. By AUGUSTUS THIERRY, Author of the "History of the Conquest of England by the Normans." 8vo., paper....................................75 cts.
Cloth $1.00

BOOK OF THE SEASONS;

Or, The Calendar of Nature. By WILLIAM HOWITT. One Volume, 12mo, cloth...$1
Calf extra..$2

PICKINGS FROM THE "PORTFOLIO OF THE REPORTER OF THE NEW ORLEANS PICAYUNE."

Comprising Sketches of the Eastern Yankee, the Western Hoosier, and such others as make up Society in the great Metropolis of the South. With Designs by DARLEY. 18mo., paper...50 cts.
Cloth..62 cts.

NOTES OF A TRAVELLER

On the Social and Political State of France, Prussia, Switzerland, Italy, and other parts of Europe, during the present Century. By SAMUEL LAING. In One Volume, 8vo., cloth.........$1

HISTORY OF THE CAPTIVITY OF NAPOLEON AT ST. HELENA.

By GENERAL COUNT MONTHOLON, the Emperor's Companion in Exile and Testamentary Executor. One Volume, 8vo., cloth, $2.50
Half morocco..$3.00

MY SHOOTING BOX.

By FRANK FORRESTER, (Henry Wm. Herbert, Esq.,) Author of "Warwick Woodlands," &c. With Illustrations, by DARLEY.
One Volume, 12mo., cloth...62 cts.
Paper covers...50 cts.

MYSTERIES OF THE BACKWOODS:

Or, Sketches of the South-west—including Character, Scenery, and Rural Sports. By T. B. THORPE, Author of "Tom Owen, the Bee-Hunter," &c. Illustrated by DARLEY. 12mo, cloth, 62 cts.
Paper...50 cts.

NARRATIVE OF THE LATE EXPEDITION TO THE DEAD SEA.

From a Diary by one of the Party. Edited by EDWARD P. MONTAGUE. 12mo, cloth...$1

MY DREAMS:

A Collection of Poems. By Mrs. LOUISA S. MCCORD. 12mo, boards ....75 cts

---

AMERICAN COMEDIES.

By JAMES K. PAULDING and WM. IRVING PAULDING. One Volume, 16mo, boards..50 cts.

RAMBLES IN YUCATAN;

Or, Notes of Travel through the Peninsula: including a Visit to the Remarkable Ruins of Chi-chen, Kabah, Zayi, and Uxmal. With numerous Illustrations. By B. M. NORMAN. Seventh Edition. In One Volume, octavo, cloth.................................$2

THE AMERICAN IN PARIS.

By JOHN SANDERSON. A New Edition. In Two Volumes, 12mo, cloth...$1

This is the most animated, graceful, and intelligent sketch of French manners, or any other, that we have had for these twenty years.—*London Monthly Magazine.*

ROBINSON CRUSOE.

A Complete Edition, with Six Illustrations. One Volume, 8vo, paper covers...$1.00
Cloth, gilt edges..$1.25

SCENES IN THE ROCKY MOUNTAINS,

And in Oregon, California, New Mexico, Texas, and the Grand Prairies; or, Notes by the Way. By RUFUS B. SAGE. Second Edition. One Volume, 12mo, paper covers ....50 cts.
With a Map, bound in cloth.......................................75 cts.

THE PUBLIC MEN OF THE REVOLUTION:

Including Events from the Peace of 1783 to the Peace of 1815. In a Series of Letters. By the late Hon. WM. SULLIVAN, LL. D. With a Biographical Sketch of the Author, by his son, JOHN T. S. SULLIVAN. With a Portrait. In One Volume, 8vo, cloth...$2

ACHIEVEMENTS OF THE KNIGHTS OF MALTA.

By ALEXANDER SUTHERLAND. In One Volume, 16mo, cloth, $1.00
Paper...75 cts.

ATALANTIS.

A Poem. By WILLIAM GILMORE SIMMS. 12mo, boards, 37 cts.

LIVES OF MEN OF LETTERS AND SCIENCE.

By Henry Lord Brougham. Two Volumes, 12mo, cloth, $1.50
Paper $1.00

THE LIFE, LETTERS, AND JOURNALS OF LORD BYRON.

By Thomas Moore. Two Volumes, 12mo, cloth...............$2

THE BOWL OF PUNCH.

Illustrated by Numerous Plates. 12mo, paper............50 cts.

CHILDREN IN THE WOOD.

Illustrated by Harvey. 12mo, cloth, gilt...................50 cts.
Paper ....25 cts.

CHARCOAL SKETCHES.

By Joseph C. Neal. With Illustrations. 12mo, paper, 25 cts.

THE POEMS OF C. P. CRANCH.

In One Volume, 12mo, boards.................................37 cts.

THE WORKS OF BENJ. DISRAELI.

Two Volumes, 8vo, cloth...$2
Paper covers...$1

NATURE DISPLAYED IN HER MODE OF TEACHING FRENCH.

By N. G. Dufief. Two Volumes, 8vo, boards..................$5

NATURE DISPLAYED IN HER MODE OF TEACHING SPANISH.

By N. G. Dufief. In Two Volumes, 8vo, boards..............$7

FRENCH AND ENGLISH DICTIONARY.

By N. G. DUFIEF. In One Volume, 8vo, sheep..................$5

FROISSART BALLADS AND OTHER POEMS.

By PHILIP PENDLETON COOKE. In One Volume, 12mo, boards...50 cts.

THE LIFE OF RICHARD THE THIRD.

By MISS HALSTED. In One Volume, 8vo, cloth...........$1.50

THE LIFE OF NAPOLEON BONAPARTE.

By WILLIAM HAZLITT. In Three Volumes, 12mo, cloth......$3
Half calf..$4

TRAVELS IN GERMANY, BY W. HOWITT.
EYRE'S NARRATIVE. BURNE'S CABOOL.

In One Volume, 8vo, cloth....................................$1.25

STUDENT-LIFE IN GERMANY.

By WILLIAM HOWITT. In One Volume, 8vo, cloth............$2

IMAGE OF HIS FATHER.

By MAYHEW. Complete in One Volume, 8vo, paper....25 cts.

SPECIMENS OF THE BRITISH CRITICS.

By CHRISTOPHER NORTH (Professor Wilson). 12mo, cloth.75 cts.

A TOUR TO THE RIVER SAUGENAY, IN LOWER CANADA.

By CHARLES LANMAN. In One Volume, 16mo, cloth....62 cts
Paper..50 cts

TRAVELS IN AUSTRIA, RUSSIA, SCOTLAND, ENGLAND AND WALES.

By J. G. Kohl. One Volume, 8vo. cloth....................$1.25

LIFE OF OLIVER GOLDSMITH.

By James Prior. In One Volume, 8vo, boards............... .2

OUR ARMY AT MONTEREY.

By T. B. Thorpe. 16mo, cloth.................................62 cts.
Paper covers...50 cts.

OUR ARMY ON THE RIO GRANDE.

By T. B. Thorpe. 16mo, cloth.................................62 cts.
Paper covers...50 cts.

LIFE OF LORENZO DE MEDICI.

By William Roscoe. In Two Volumes, 8vo, cloth...........$3

MISCELLANEOUS ESSAYS OF SIR WALTER SCOTT.

In Three Volumes, 12mo, cloth.................................$3.50
Half morocco..$4.25

SERMON ON THE MOUNT.

Illuminated. Boards..$1.50
" Silk..$2.00
" Morocco super.....................................$3.00

MISCELLANEOUS ESSAYS OF THE REV. SYDNEY SMITH.

In Three Volumes, 12mo, cloth.................................$3.50
Half morocco..$4.25

MRS. CAUDLE'S CURTAIN LECTURES................12½ cts.

SERMONS BY THE REV. SYDNEY SMITH.

One Volume, 12mo, cloth..75 cts.

MISCELLANEOUS ESSAYS OF SIR JAMES STEPHEN.

One Volume, 12mo, cloth..$1.25

THREE HOURS; OR, THE VIGIL OF LOVE.

A Volume of Poems. By Mrs. Hale. 18mo, boards.. 75 cts

TORLOGH O'BRIEN:

A Tale of the Wars of King James. 8vo, paper covers 12½ cts.
Illustrated..37½ cts.

AN AUTHOR'S MIND.

Edited by M. F. Tupper. One Volume, 16mo, cloth....62 cts.
Paper covers..50 cts.

HISTORY OF THE ANGLO-SAXONS.

By Sharon Turner. Two Volumes, 8vo, cloth............$4.50

PROSE WORKS OF N. PARKER WILLIS.

In One Volume, 8vo, 800 pp., cloth, gilt......................$3.00
Cloth extra, gilt edges..$3.50
Library sheep...$3.50
Turkey morocco backs...$3.75
" extra...$5.50

MISCELLANEOUS ESSAYS OF PROF. WILSON.

Three Volumes, 12mo, cloth.....................................$3.50

WORD TO WOMAN.

By Caroline Fry. 12mo, cloth...............................60 cts.

WYATT'S HISTORY OF THE KINGS OF FRANCE.

Illustrated by 72 Portraits. One Volume, 16mo, cloth...$1.00
Cloth, extra gilt...$1.25

www.ingramcontent.com/pod-product-compliance
Lightning Source LLC
LaVergne TN
LVHW010253110826
845151LV00004B/1457

* 9 7 8 1 4 2 5 5 2 2 6 7 4 *